Fossils: A Very Short Introduction

Very Short Introductions available now:

ACCOUNTING Christopher Nobes
ADVERTISING Winston Fletcher
AFRICAN HISTORY John Parker and Richard Rathbone
AGNOSTICISM Robin Le Poidevin
ALEXANDER THE GREAT Hugh Bowden
AMERICAN HISTORY Paul S. Boyer
AMERICAN IMMIGRATION David A. Gerber
AMERICAN POLITICAL PARTIES AND ELECTIONS L. Sandy Maisel
AMERICAN POLITICS Richard M. Valelly
THE AMERICAN PRESIDENCY Charles O. Jones
ANAESTHESIA Aidan O'Donnell
ANARCHISM Colin Ward
ANCIENT EGYPT Ian Shaw
ANCIENT GREECE Paul Cartledge
THE ANCIENT NEAR EAST Amanda H. Podany
ANCIENT PHILOSOPHY Julia Annas
ANCIENT WARFARE Harry Sidebottom
ANGELS David Albert Jones
ANGLICANISM Mark Chapman
THE ANGLO-SAXON AGE John Blair
THE ANIMAL KINGDOM Peter Holland
ANIMAL RIGHTS David DeGrazia
THE ANTARCTIC Klaus Dodds
ANTISEMITISM Steven Beller
ANXIETY Daniel Freeman and Jason Freeman
THE APOCRYPHAL GOSPELS Paul Foster
ARCHAEOLOGY Paul Bahn
ARCHITECTURE Andrew Ballantyne
ARISTOCRACY William Doyle
ARISTOTLE Jonathan Barnes
ART HISTORY Dana Arnold
ART THEORY Cynthia Freeland
ASTROBIOLOGY David C. Catling
ATHEISM Julian Baggini
AUGUSTINE Henry Chadwick
AUSTRALIA Kenneth Morgan
AUTISM Uta Frith
THE AVANT GARDE David Cottington
THE AZTECS Davíd Carrasco
BACTERIA Sebastian G. B. Amyes
BARTHES Jonathan Culler
THE BEATS David Sterritt
BEAUTY Roger Scruton
BESTSELLERS John Sutherland
THE BIBLE John Riches
BIBLICAL ARCHAEOLOGY Eric H. Cline
BIOGRAPHY Hermione Lee
THE BLUES Elijah Wald
THE BOOK OF MORMON Terryl Givens
BORDERS Alexander C. Diener and Joshua Hagen
THE BRAIN Michael O'Shea
THE BRITISH CONSTITUTION Martin Loughlin
THE BRITISH EMPIRE Ashley Jackson
BRITISH POLITICS Anthony Wright
BUDDHA Michael Carrithers
BUDDHISM Damien Keown
BUDDHIST ETHICS Damien Keown
CANCER Nicholas James
CAPITALISM James Fulcher
CATHOLICISM Gerald O'Collins
CAUSATION Stephen Mumford and Rani Lill Anjum
THE CELL Terence Allen and Graham Cowling
THE CELTS Barry Cunliffe
CHAOS Leonard Smith
CHILDREN'S LITERATURE Kimberley Reynolds
CHINESE LITERATURE Sabina Knight
CHOICE THEORY Michael Allingham
CHRISTIAN ART Beth Williamson
CHRISTIAN ETHICS D. Stephen Long
CHRISTIANITY Linda Woodhead
CITIZENSHIP Richard Bellamy
CIVIL ENGINEERING David Muir Wood
CLASSICAL LITERATURE William Allan
CLASSICAL MYTHOLOGY Helen Morales
CLASSICS Mary Beard and John Henderson
CLAUSEWITZ Michael Howard
CLIMATE Mark Maslin
THE COLD WAR Robert McMahon
COLONIAL AMERICA Alan Taylor
COLONIAL LATIN AMERICAN LITERATURE Rolena Adorno
COMEDY Matthew Bevis
COMMUNISM Leslie Holmes
COMPLEXITY John H. Holland
THE COMPUTER Darrel Ince
THE CONQUISTADORS Matthew Restall and Felipe Fernández-Armesto
CONSCIENCE Paul Strohm
CONSCIOUSNESS Susan Blackmore
CONTEMPORARY ART Julian Stallabrass
CONTEMPORARY FICTION Robert Eaglestone
CONTINENTAL PHILOSOPHY Simon Critchley
CORAL REEFS Charles Sheppard
COSMOLOGY Peter Coles
CRITICAL THEORY Stephen Eric Bronner
THE CRUSADES Christopher Tyerman
CRYPTOGRAPHY Fred Piper and Sean Murphy
THE CULTURAL REVOLUTION Richard Curt Kraus
DADA AND SURREALISM David Hopkins
DARWIN Jonathan Howard
THE DEAD SEA SCROLLS Timothy Lim
DEMOCRACY Bernard Crick
DERRIDA Simon Glendinning
DESCARTES Tom Sorell
DESERTS Nick Middleton
DESIGN John Heskett
DEVELOPMENTAL BIOLOGY Lewis Wolpert
THE DEVIL Darren Oldridge
DIASPORA Kevin Kenny
DICTIONARIES Lynda Mugglestone
DINOSAURS David Norman
DIPLOMACY Joseph M. Siracusa
DOCUMENTARY FILM Patricia Aufderheide
DREAMING J. Allan Hobson
DRUGS Leslie Iversen
DRUIDS Barry Cunliffe
EARLY MUSIC Thomas Forrest Kelly
THE EARTH Martin Redfern
ECONOMICS Partha Dasgupta
EDUCATION Gary Thomas
EGYPTIAN MYTH Geraldine Pinch
EIGHTEENTH-CENTURY BRITAIN Paul Langford
THE ELEMENTS Philip Ball
EMOTION Dylan Evans
EMPIRE Stephen Howe
ENGELS Terrell Carver
ENGINEERING David Blockley
ENGLISH LITERATURE Jonathan Bate
ENVIRONMENTAL ECONOMICS Stephen Smith
EPIDEMIOLOGY Rodolfo Saracci
ETHICS Simon Blackburn

THE EUROPEAN UNION John Pinder and Simon Usherwood
EVOLUTION Brian and Deborah Charlesworth
EXISTENTIALISM Thomas Flynn
THE EYE Michael Land
FAMILY LAW Jonathan Herring
FASCISM Kevin Passmore
FASHION Rebecca Arnold
FEMINISM Margaret Walters
FILM Michael Wood
FILM MUSIC Kathryn Kalinak
THE FIRST WORLD WAR Michael Howard
FOLK MUSIC Mark Slobin
FOOD John Krebs
FORENSIC PSYCHOLOGY David Canter
FORENSIC SCIENCE Jim Fraser
FOSSILS Keith Thomson
FOUCAULT Gary Gutting
FRACTALS Kenneth Falconer
FREE SPEECH Nigel Warburton
FREE WILL Thomas Pink
FRENCH LITERATURE John D. Lyons
THE FRENCH REVOLUTION William Doyle
FREUD Anthony Storr
FUNDAMENTALISM Malise Ruthven
GALAXIES John Gribbin
GALILEO Stillman Drake
GAME THEORY Ken Binmore
GANDHI Bhikhu Parekh
GENIUS Andrew Robinson
GEOGRAPHY John Matthews and David Herbert
GEOPOLITICS Klaus Dodds
GERMAN LITERATURE Nicholas Boyle
GERMAN PHILOSOPHY Andrew Bowie
GLOBAL CATASTROPHES Bill McGuire
GLOBAL ECONOMIC HISTORY Robert C. Allen
GLOBAL WARMING Mark Maslin
GLOBALIZATION Manfred Steger
THE GOTHIC Nick Groom
GOVERNANCE Mark Bevir
THE GREAT DEPRESSION AND THE NEW DEAL Eric Rauchway
HABERMAS James Gordon Finlayson
HAPPINESS Daniel M. Haybron
HEGEL Peter Singer
HEIDEGGER Michael Inwood
HERODOTUS Jennifer T. Roberts
HIEROGLYPHS Penelope Wilson
HINDUISM Kim Knott
HISTORY John H. Arnold
THE HISTORY OF ASTRONOMY Michael Hoskin
THE HISTORY OF LIFE Michael Benton
THE HISTORY OF MATHEMATICS Jacqueline Stedall
THE HISTORY OF MEDICINE William Bynum
THE HISTORY OF TIME Leofranc Holford-Strevens
HIV/AIDS Alan Whiteside
HOBBES Richard Tuck
HORMONES Martin Luck
HUMAN EVOLUTION Bernard Wood
HUMAN RIGHTS Andrew Clapham
HUMANISM Stephen Law
HUME A. J. Ayer
HUMOUR Noël Carroll
THE ICE AGE Jamie Woodward
IDEOLOGY Michael Freeden
INDIAN PHILOSOPHY Sue Hamilton
INFORMATION Luciano Floridi
INNOVATION Mark Dodgson and David Gann
INTELLIGENCE Ian J. Deary
INTERNATIONAL MIGRATION Khalid Koser
INTERNATIONAL RELATIONS Paul Wilkinson
INTERNATIONAL SECURITY Christopher S. Browning
ISLAM Malise Ruthven
ISLAMIC HISTORY Adam Silverstein
ITALIAN LITERATURE Peter Hainsworth and David Robey
JESUS Richard Bauckham
JOURNALISM Ian Hargreaves
JUDAISM Norman Solomon
JUNG Anthony Stevens
KABBALAH Joseph Dan
KAFKA Ritchie Robertson
KANT Roger Scruton
KEYNES Robert Skidelsky
KIERKEGAARD Patrick Gardiner
THE KORAN Michael Cook
LANDSCAPE ARCHITECTURE Ian H. Thompson
LANDSCAPES AND GEOMORPHOLOGY Andrew Goudie and Heather Viles
LANGUAGES Stephen R. Anderson
LATE ANTIQUITY Gillian Clark
LAW Raymond Wacks
THE LAWS OF THERMODYNAMICS Peter Atkins
LEADERSHIP Keith Grint
LINCOLN Allen C. Guelzo
LINGUISTICS Peter Matthews
LITERARY THEORY Jonathan Culler
LOCKE John Dunn
LOGIC Graham Priest
MACHIAVELLI Quentin Skinner
MADNESS Andrew Scull
MAGIC Owen Davies
MAGNA CARTA Nicholas Vincent
MAGNETISM Stephen Blundell
MALTHUS Donald Winch
MANAGEMENT John Hendry
MAO Delia Davin
MARINE BIOLOGY Philip V. Mladenov
THE MARQUIS DE SADE John Phillips
MARTIN LUTHER Scott H. Hendrix
MARTYRDOM Jolyon Mitchell
MARX Peter Singer
MATHEMATICS Timothy Gowers
THE MEANING OF LIFE Terry Eagleton
MEDICAL ETHICS Tony Hope
MEDICAL LAW Charles Foster
MEDIEVAL BRITAIN John Gillingham and Ralph A. Griffiths
MEMORY Jonathan K. Foster
METAPHYSICS Stephen Mumford
MICHAEL FARADAY Frank A.J.L. James
MICROECONOMICS Avinash Dixit
MODERN ART David Cottington
MODERN CHINA Rana Mitter
MODERN FRANCE Vanessa R. Schwartz
MODERN IRELAND Senia Pašeta
MODERN JAPAN Christopher Goto-Jones
MODERN LATIN AMERICAN LITERATURE Roberto González Echevarría
MODERN WAR Richard English
MODERNISM Christopher Butler
MOLECULES Philip Ball
THE MONGOLS Morris Rossabi
MORMONISM Richard Lyman Bushman
MUHAMMAD Jonathan A.C. Brown

MULTICULTURALISM Ali Rattansi
MUSIC Nicholas Cook
MYTH Robert A. Segal
THE NAPOLEONIC WARS Mike Rapport
NATIONALISM Steven Grosby
NELSON MANDELA Elleke Boehmer
NEOLIBERALISM Manfred Steger and Ravi Roy
NETWORKS Guido Caldarelli and Michele Catanzaro
THE NEW TESTAMENT Luke Timothy Johnson
THE NEW TESTAMENT AS LITERATURE Kyle Keefer
NEWTON Robert Iliffe
NIETZSCHE Michael Tanner
NINETEENTH-CENTURY BRITAIN Christopher Harvie and H. C. G. Matthew
THE NORMAN CONQUEST George Garnett
NORTH AMERICAN INDIANS Theda Perdue and Michael D. Green
NORTHERN IRELAND Marc Mulholland
NOTHING Frank Close
NUCLEAR POWER Maxwell Irvine
NUCLEAR WEAPONS Joseph M. Siracusa
NUMBERS Peter M. Higgins
NUTRITION David A. Bender
OBJECTIVITY Stephen Gaukroger
THE OLD TESTAMENT Michael D. Coogan
THE ORCHESTRA D. Kern Holoman
ORGANIZATIONS Mary Jo Hatch
PAGANISM Owen Davies
THE PALESTINIAN-ISRAELI CONFLICT Martin Bunton
PARTICLE PHYSICS Frank Close
PAUL E. P. Sanders
PENTECOSTALISM William K. Kay
THE PERIODIC TABLE Eric R. Scerri
PHILOSOPHY Edward Craig
PHILOSOPHY OF LAW Raymond Wacks
PHILOSOPHY OF SCIENCE Samir Okasha
PHOTOGRAPHY Steve Edwards
PLAGUE Paul Slack
PLANETS David A. Rothery
PLANTS Timothy Walker
PLATO Julia Annas
POLITICAL PHILOSOPHY David Miller
POLITICS Kenneth Minogue
POSTCOLONIALISM Robert Young
POSTMODERNISM Christopher Butler
POSTSTRUCTURALISM Catherine Belsey
PREHISTORY Chris Gosden
PRESOCRATIC PHILOSOPHY Catherine Osborne
PRIVACY Raymond Wacks
PROBABILITY John Haigh
PROGRESSIVISM Walter Nugent
PROTESTANTISM Mark A. Noll
PSYCHIATRY Tom Burns
PSYCHOLOGY Gillian Butler and Freda McManus
PURITANISM Francis J. Bremer
THE QUAKERS Pink Dandelion
QUANTUM THEORY John Polkinghorne
RACISM Ali Rattansi
RADIOACTIVITY Claudio Tuniz
RASTAFARI Ennis B. Edmonds
THE REAGAN REVOLUTION Gil Troy
REALITY Jan Westerhoff
THE REFORMATION Peter Marshall
RELATIVITY Russell Stannard
RELIGION IN AMERICA Timothy Beal
THE RENAISSANCE Jerry Brotton
RENAISSANCE ART Geraldine A. Johnson
REVOLUTIONS Jack A. Goldstone
RHETORIC Richard Toye
RISK Baruch Fischhoff and John Kadvany
RIVERS Nick Middleton
ROBOTICS Alan Winfield
ROMAN BRITAIN Peter Salway
THE ROMAN EMPIRE Christopher Kelly
THE ROMAN REPUBLIC David M. Gwynn
ROMANTICISM Michael Ferber
ROUSSEAU Robert Wokler
RUSSELL A. C. Grayling
RUSSIAN HISTORY Geoffrey Hosking
RUSSIAN LITERATURE Catriona Kelly
THE RUSSIAN REVOLUTION S. A. Smith
SCHIZOPHRENIA Chris Frith and Eve Johnstone
SCHOPENHAUER Christopher Janaway
SCIENCE AND RELIGION Thomas Dixon
SCIENCE FICTION David Seed
THE SCIENTIFIC REVOLUTION Lawrence M. Principe
SCOTLAND Rab Houston
SEXUALITY Véronique Mottier
SHAKESPEARE Germaine Greer
SIKHISM Eleanor Nesbitt
THE SILK ROAD James A. Millward
SLEEP Steven W. Lockley and Russell G. Foster
SOCIAL AND CULTURAL ANTHROPOLOGY John Monaghan and Peter Just
SOCIALISM Michael Newman
SOCIOLINGUISTICS John Edwards
SOCIOLOGY Steve Bruce
SOCRATES C. C. W. Taylor
THE SOVIET UNION Stephen Lovell
THE SPANISH CIVIL WAR Helen Graham
SPANISH LITERATURE Jo Labanyi
SPINOZA Roger Scruton
SPIRITUALITY Philip Sheldrake
STARS Andrew King
STATISTICS David J. Hand
STEM CELLS Jonathan Slack
STUART BRITAIN John Morrill
SUPERCONDUCTIVITY Stephen Blundell
SYMMETRY Ian Stewart
TEETH Peter S. Ungar
TERRORISM Charles Townshend
THEOLOGY David F. Ford
THOMAS AQUINAS Fergus Kerr
THOUGHT Tim Bayne
TIBETAN BUDDHISM Matthew T. Kapstein
TOCQUEVILLE Harvey C. Mansfield
TRAGEDY Adrian Poole
THE TROJAN WAR Eric H. Cline
TRUST Katherine Hawley
THE TUDORS John Guy
TWENTIETH-CENTURY BRITAIN Kenneth O. Morgan
THE UNITED NATIONS Jussi M. Hanhimäki
THE U.S. CONGRESS Donald A. Ritchie
THE U.S. SUPREME COURT Linda Greenhouse
UTOPIANISM Lyman Tower Sargent
THE VIKINGS Julian Richards
VIRUSES Dorothy H. Crawford
WITCHCRAFT Malcolm Gaskill
WITTGENSTEIN A. C. Grayling
WORK Stephen Fineman
WORLD MUSIC Philip Bohlman
THE WORLD TRADE ORGANIZATION Amrita Narlikar
WRITING AND SCRIPT Andrew Robinson

Keith Thomson

FOSSILS

A Very Short Introduction

OXFORD
UNIVERSITY PRESS

OXFORD
UNIVERSITY PRESS

Great Clarendon Street, Oxford OX2 6DP

Oxford University Press is a department of the University of Oxford.
It furthers the University's objective of excellence in research, scholarship,
and education by publishing worldwide in

Oxford New York

Auckland Cape Town Dar es Salaam Hong Kong Karachi
Kuala Lumpur Madrid Melbourne Mexico City Nairobi
New Delhi Shanghai Taipei Toronto

With offices in

Argentina Austria Brazil Chile Czech Republic France Greece
Guatemala Hungary Italy Japan Poland Portugal Singapore
South Korea Switzerland Thailand Turkey Ukraine Vietnam

Oxford is a registered trade mark of Oxford University Press
in the UK and in certain other countries

Published in the United States
by Oxford University Press Inc., New York

First published as a Very Short Introduction 2005

British Library Cataloguing in Publication Data
Data available

Library of Congress Cataloging in Publication Data
Data available

ISBN-13: 978-0-19-280504-1

Typeset by RefineCatch Ltd, Bungay, Suffolk

Printed and bound by
CPI Group (UK) Ltd, Croydon, CR0 4YY

Contents

Acknowledgements

As an evolutionary biologist with interests in development and physiology, the attraction of fossils for me has been twofold: to discover how they illuminate our ideas about evolution, and to find ways of using our knowledge of living organisms to make fossils come 'alive'. Although I have spent more time than I care to remember on working with fossils, I did not set out to be a palaeontologist. I am particularly grateful, therefore, to my colleagues on both sides of the scholarly neontological/palaeontological coin for tolerating my invasions, over the years, into their territories and even assisting me in the process. I have always worked with vertebrate fossils, rather than invertebrates, plants, or fungi, and that bias shows in the examples I use; the principles, however, are common to all fossils.

I must thank Erik Sperling for invaluable research assistance and Marsha Filion at Oxford University Press for her enthusiastic encouragement. Linda Price Thomson, Jim Kennedy, Kristin Andrews-Speed, Mark Sutton, Ian Tattersall, Gino Segre, and Anthony Fiorillo kindly read all or part of the manuscript and smoothed over the rough patches. Eliza Howlett, Derek Siveter, Philip Powell, Mark Robinson, Bethia Thomas, Dinah Birch, Ted Daeschler, and Carl Thompson also made invaluable contributions. Linda Price Thomson drew Figures 14, 18, and 21.

List of illustrations

The publisher and the author apologize for any errors or omissions in the above list. If contacted they will be pleased to rectify these at the earliest opportunity.

Chapter 1
Introduction

Latin *fossilis*: dug up.

I vividly remember when and where I found my first fossil. It was early April 1961, and the place was Archer County, Texas, then, as now, a hardscrabble sort of a landscape, dry and dissected by shallow washes where the grey-green and red Permian rocks are exposed and where rattlesnakes thrive. Fossils have been found in these rocks for over a hundred years. We were searching for fishes, early amphibians, and reptiles, and my first find was a single grey vertebra. Under the encrusting lime, the canal for the spinal cord was visible, together with the facets for articulation with adjacent vertebrae. Exploration on hands and knees revealed other bits and pieces, all from the tail of a crocodile-sized amphibian called *Eryops*. The animal had probably died somewhere else, as there were no other remains; these few bones had been washed downstream and deposited in a shallow lens of silt. Silt and bones had then been buried under more layers of sediment and slowly transformed into rock. That had been 220 million years ago when the region was a marshy river delta. Other fossil-bearing pockets nearby contained fish scales and shark spines. Some contained the remains of the extraordinary *Dimetrodon* – a reptile with the spines of its vertebrae extended to form a high sail on its back. In pure scientific terms, my first fossil was not nearly as interesting. But I was hooked.

In this first paragraph I have made some statements of fact (the existence of the fossil; its shape; the identity of the animal it came from; its petrified nature; the associated remains) and some inferences from other facts (the age of the rocks; what happened to the original animal when it died; the original environment where this all happened). In this book I will explain the basis for all that: what fossils are and some of the concepts and principles upon which the study of fossils is based. I will discuss also the broader significance of fossils in teaching us about the history of the earth and the animals and plants – including our own ancestors – that have variously inhabited it for the past few billion years.

Since antiquity, explanations of what fossils are and theories of what they mean have had a varied history. At first, the word had been used for anything dug up from the earth, including minerals, gems, or metal ores, as well as the petrified organic remains to which we now restrict the term. Classical Greek authors such as Empedocles and Xenophanes had a pretty good idea of what fossils were, as had Leonardo da Vinci, but fossils became especially important when all the intersecting philosophical/scientific consequences of the very existence of fossils in the earth reached a critical point. We can even pinpoint the author and the date: the English scientist Robert Hooke, writing in 1665. Before then, fossils could be treated as curiosities; since then, fossils have become variously the foundation of a scientific revolution and a threat to the fundamentals of theology.

Before Hooke, fossils could be dismissed as mere 'sports of nature' – 'formed stones' – and elaborate theories had to be dreamt up to explain them in terms of a 'Plastick Virtue' in the soil or the properties of crystals. For others, fossils were the physical evidence of the great biblical Flood. But for the scientist, fossils became the central facts of a theory of a changing earth of great antiquity. They led us to understand the restless movements of continents, fluctuating climates, and a history of life undergoing inexorable processes of origination and extinction.

1. Robert Hooke's accurate drawings of fossils, as in this plate of ammonites from his *Lectures and Discourses of Earthquakes* (published posthumously in 1705), helped convince readers of their organic nature

By studying fossils, we can detect changing patterns in the diversity of life on earth, discovering that there have been sudden periods of mass extinction, others of strong diversification. Fossils help show how the continental plates have drifted around the surface of the earth and how the surface of the earth has changed; they show, for example, that deep seas once lay where there is now dry land. We can chart ancient changes in climate, discovering among other things that the present Arctic and Antarctic were once subtropical paradises.

Fossils had started to prove all this long before Charles Darwin's theory of natural selection, formally proposed in 1859, provided the causal mechanism for the origin of species. Fossils of the reptile/bird *Archaeopteryx* (1860) and Neanderthal man (1856) were discovered just in time to give substance to his theories: they were 'missing links' in a continuous chain of existence reaching back to the beginning of life. Now, every new discovery redefines our search for new 'links'; we are on the search for fishes with legs, dinosaurs with feathers, and, always, for human ancestors. With respect to human evolution, just as Galileo with his telescope revealed the existence of worlds beyond worlds out there in space and thus reduced the earth (and man) to an insignificant speck in the cosmos, the history of fossils in this very old earth exposes *Homo sapiens* as simply a Johnny-come-lately in the animal world, and a creature most likely doomed to extinction just like the rest.

Fossils provide a highly accessible kind of science. Many a serious scholar had his first interest in science triggered by an enthusiasm for fossils. Natural history museums depend on fossils, and particularly dinosaurs, for a large part of their audience and income, and they depend on fossil hunters to present the subject to the public. For many palaeontologists, professional or amateur, fossils represent a happy fusion between the romanticism of the 19th century and the cold, hard clarity of contemporary science. Fossil collecting, whether out on some vast foreign plain, or scrambling among the cliff falls at Lyme Regis, remains one of the very few activities (amateur astronomy is another) whereby a person working alone, or in a small group, can accomplish great things. Armed only with a hammer and a good eye, like a prospector for gold, he or she can make a fundamental contribution to science.

Both amateur and professional palaeontology have expanded enormously in the past 50 years. When I first attended a meeting of the Society of Vertebrate Palaeontology in 1961, there were about 30 members present. Last year there were more than 2,000.

Creatures such as dinosaurs, ammonites, trilobites, flying reptiles, and mammoths (fossil plants rarely enter the public imagination) are half real and half unreal. We are fascinated equally by their familiarity and their foreignness. They may even be cuddly, for example for the 6-year-old who has already mastered the tongue-twisting lexicon of their Latin names, and soon will be collecting accurately modelled replicas to go with the soft toys he keeps in the bedroom – thereby supporting a vast industry.

While dinosaurs belong in the distant past, *Homo erectus* and *Homo Neanderthalensis*, on the other hand, are faintly alarming; in every sense being far too close for comfort. We do not have to resort to lurid, far-fetched caricatures of our predecessors and cousins as shambling, hairy brutes to accept that, only a vanishingly short time ago, as measured in the geological frame, our forebears were without language or material culture. A fossil record that says that painting and carving arose only some 30,000 to 40,000 years ago, and within a people who were physically identical to us, either makes us feel especially ennobled by whatever triggered the origin of technology and a culture that has given us Rembrandt, Turner, Twyla Tharp, the Beatles, and Shakespeare, or it leaves us totally humbled. No wonder then that the idea that we humans were specially created by God has its attractions.

But dinosaurs and humans are only the two components of a vast spectrum of fossil life. Stretching back almost to the beginnings of the earth are literally hundreds of thousands of species represented by countless millions of apparently unprepossessing specimens lying, dead as the dodo, in museum collections (and huge numbers more that are still buried in the rock). This is where scientists genuinely wearing white lab coats come to the fore. They can count, measure, dissect, X-ray or CAT scan, or model by computer, and then build views of the world that we could otherwise only dream of. They can both document the course of evolutionary change and lead us to views of possible mechanisms. Examination of a golf-ball-sized chunk of fossil sea bed can tell us where to drill for oil or gas. Fossils too small to be seen with the naked eye tell us that 700 million years ago the earth was buried in an ice age far greater than the last one; they can also tell us about more recent climates and, in the process, warn us about the future.

Every day, somewhere around the earth, dozens of palaeontologists are digging somewhere new, or scavenging old deposits and museum collections, for yet another fragment of insight into the earth and life sciences. And there is a great deal still to learn about fossils themselves and about the vagaries of their dying that allowed (against enormous odds) some individuals to be preserved and turned into rock. Also, because fossils are so much in the public eye, there are always new fakes to be unmasked and false theories to be rejected. And magnificent discoveries still to be made. Simply by digging in the ground.

Chapter 2
A cultural phenomenon

There is something intriguing about a whole discipline founded on organisms that have become important to us only in, and by means of, their death. Fossils fascinate us both when they are most different from modern life on earth, and are separated from us by time intervals that are almost unimaginable, and when they link living species such as ourselves to our immediate forebears. From whatever age, those dead organisms that lived in other times are both quite unreal to us, and at the same time strangely familiar. Fossils reveal to us ancient worlds populated by strange beasts and weird plants, whose existences were curiously like and yet fascinatingly different from our present world. They not only capture our imagination, they test our ideas about life itself. Indeed, it is impossible to imagine what our present view of the world and ourselves would be if we had never known about fossils at all.

Fossils before the Enlightenment

Although general public acceptance of the organic nature of fossils – that they are the remains of once-living organisms, preserved in and themselves transformed to rock – did not come until the turn of the 19th century, modern palaeontology began in the last third of the 17th century with the writings of Robert Hooke (in *Micrographia*, 1665, and *Discourse of Earthquakes*, 1668), followed in 1669 by the *Prodromus* of Niels Stensen (later Nicolai Stenonis and now always known simply as Steno). Hooke was a true genius and polymath at the Royal Society in London who seems to have studied geology very informally. No less brilliant, Steno was first an anatomist in Leiden and then in the Medici court at Florence. He devoted years of study of the geology of Tuscany before adopting a life of self-denial as a Catholic priest and bishop.

Before Hooke and Steno, explanations of the nature and causes of fossils exercised philosophers of all kinds. An early obstacle to unlocking the secrets of fossils was that they seemed easiest to find in cliffs and mountains. If they were the remains of real fish and clams, how on earth (so to speak) did they get there? It did not seem possible that the earth could have been so changed that what was once the bottom of the sea is now thousands of metres in the air. Leonardo da Vinci offered what seemed the only possible solution: that sea levels had dropped. A similar explanation was offered by Steno. Hooke, on the other hand, insisted that mountains were raised up from the sea floor by earthquakes and the earth's 'inner heat'. Without the benefit of an advanced understanding of the gigantic forces that (usually) imperceptibly shape and change the earth, and of the immense expanse of geological time, such explanations seemed at best far-fetched.

Another difficulty was that fossil creatures were notably different from living ones. Were they faulty versions of modern species or bizarre 'aberrations from nature'? The concept of *extinction* was obvious to Hooke, but it squarely opposed the biblical account of Creation which speaks of a single creating event. Extinction implied that there had been more than one episode of Creation and that, in allowing those creatures to become extinct, God had, as it were, changed his mind or even admitted to mistakes.

Recognition that the earth's crust contains multiple layers of rocks, thousands of feet thick, containing diverse fossil assemblages (mostly deposited under water), forced scholars to face the issue of mountain-building and other drastic rearrangements of the earth's

Fossils on mountains

Now if all these Bodies have been really such Shells of Fishes as they most resemble, and that these are found at the tops of the most considerable Mountains in the World . . . 'tis a very cogent Argument that the superficial Parts of the earth have been very much changed since the beginning, that the tops of Mountains have been under the Water, and arguably also, that divers parts of the bottom of the Sea have been heretofore Mountains.

Robert Hooke, *Discourse of Earthquakes* (1668)

surface. If those fossils were once living in the sea and were deposited in marine beds and are now hundreds or thousands of feet above sea level, then the earth *must* have been raised up. But the mechanisms for mountain-building remained secret. It is an extraordinary accomplishment for geology and palaeontology to have proceeded to develop and flourish while lacking such an explanation, which has only come in modern times with the discovery of the mechanisms by which vast portions of the earth's surface have been moved around over the aeons. If there had been independent, generally accepted evidence that the earth was very old and had steadily undergone changes of the sort that could thrust mountains up out of the sea, then it would have been easier to accept that fossils were true organic remains and that marine shells could be found in old rocks thousands of feet up hillsides. Equally, if there had been incontrovertible evidence that fossils were the remains of once-living organisms, then the notion of an old, changing earth would have followed more readily. In the event, understanding had to edge forwards slowly, iteratively – a discovery here, an insight there.

Philosophers also investigated the proposition that fossilization was not a natural process and fossils were not 'real' at all. First, and simplest, fossils might simply be accidents of nature – pieces of rock that merely mimic true organisms. And there is no shortage of the latter – flints shaped like a heart or a foot are easy to find in chalk deposits, for example. Alternatively, they might have been made by a God or gods who created them supernaturally, in which case those gods had also to have created all the layered rocks that contain the fossils, together with all the other apparent evidence of antiquity and change. In the biblical account in *Genesis*, this would have happened during the first days of Creation when the earth had been formed but living organisms still had not. Perhaps the extreme version of a 'Creation theory' was expounded by Philip Henry Gosse in his *Omphalos* (1856). For Gosse, a God who could make the earth and all its living creatures could easily have salted his newly minted rocks with ancient-looking fossils at the same time. As there is, and can be, no empirical evidence for such a completely *ad hoc* explanation, acceptance of it was (and is) a matter of faith rather than science, and the consequent philosophical question then became: why would any God have done that?

A quite different possibility was that fossils might be artefacts of some natural property of the rocks themselves – a process that produces mineral mimics of real organisms. Such a property was

On extinction

Certainly there are many Species of Nature that we have never seen, and there may have been also many such Species in former Ages of the World that may not be in being at present, and many variations of these Species now, which may not have had a being in former Times: We see what variety of Species, variety of Soils and Climates, and other Circumstantial Accidents do produce.

Robert Hooke, Lecture to the Royal Society, 25 July 1694

usually called a 'Plastick Virtue'. The idea depended on the proposition that, if a plant grows out of the soil, why should a fossil not grow out of the rock? While this was a popular idea in the 17th and early 18th centuries, no-one could imagine what the material nature – the actual causative element – of a 'Plastick Virtue' might be. However, there was an obvious connection to the phenomenon of crystallization, and many pseudo-fossils exist in the form of the fern-like crystallization of salts on a bedding plane.

A compromise view was that fossils developed from some kind of seeds, deposited in the rocks at Creation, which then germinated later. This would explain the fact that fossils were often found high up mountainsides. A parallel explanation was that these seeds were actually the product of living sea creatures that were dispersed to land by wind and rain, fell into crevices in the rocks, and germinated there – imperfectly so, with the result that fossil organisms are distorted rather than precise copies of living ones.

The final, and most obvious and popular, explanation of the very existence of fossils, and much of the geological condition of the earth, was Noah's Flood. Until the 1830s, the fact that most of Europe and North America is covered by thick layers of water-borne sands and gravels, with valleys carved out by water action, seemed to provide ample evidence for a great Diluvial episode. There are still those who believe, for example, that the Flood, rather than aeons of erosion by the Colorado River, created the Arizona Grand Canyon.

Many scholars followed Steno, the cleric Thomas Burnet (1681), and the physician John Woodward (1695) in believing that the biblical 'opening of the fountains of the deep' during the Flood described the earth's crust being broken like an egg, producing mountains and all the evidence we see around us of a 'broken and shattered earth'. Woodward extended the idea to the extent that the Flood then dissolved or suspended all the matter in the earth's crust and deposited it in discrete layers, according to specific gravity. In

2. Not a fossil: this mineral deposit (technical name pyrolusite, composed of manganese oxide) from the Solnhofen lithographic limestone has grown in a fern-like pattern but is definitely inorganic

all such theories, fossils represent the remains of the creatures killed in the Flood. In trying to create a material, geological explanation of fossils, such authors had to ignore points such as the prior existence of mountains in the very story they were trying to uphold, but there is no point now in refuting such theories. One difficulty that confronted contemporary scholars is worth noting, however. If, as they calculated, the pre-Flood population of the world was 8 million, and all but one family died, human fossils should be common instead of (until the discovery of Neanderthals in 1856) absent.

In fact, the record of the rocks – layer after layer, age after age – reveals multiple, overlapping, extinct worlds, each with their own characteristic organisms. Any Diluvial explanation would have to involve many, many Floods. Basically the Flood hypothesis fails because the earth's crust has not been shaped by a single event but by almost an infinity of events. Life on earth has changed over billions of years, driven by the countless 'natural shocks the flesh is heir to'. In modern theory, the earth is shaped by erosion and deposition, by earthquakes and volcanoes, and by the movement of huge areas of the earth's crust due to processes in the deeper semi-solid mantle (plate tectonics). The final nail in the coffin of the Diluvial theory was provided by Louis Agassiz, who in 1837 showed that many of the erratic boulders, water-borne sands and gravels that had seemed to be evidence for the Flood were the product of Pleistocene glacial action. The changes of flora and fauna before, after, and in between periods of glacial activity over the past 1.8 million years are due to huge climate shifts between glacial and interglacial episodes. And it eventually turned out that even humans had their ancient fossil ancestors. Fossils became the prime evidence for theories of evolution.

Fossils and philosophy

Broadly put, fossils give us an extended view of life itself, projecting life into a time dimension in which an anthropocentric viewpoint is meaningless. Whether one's view is that fossils represent the operation of natural, law-like processes, or that the whole world, including fossils, was (and still is) caused by supra-natural phenomena, fossils are always a key part of the discussion.

While some philosophical issues have long since been resolved, the basic (fundamental, as one might indeed say) problem for most religious viewpoints has not gone away. Simply put, the testimony of the rocks (to subvert the title of an 1857 book by God-fearing Hugh Miller) contradicts the account of a single act of Creation given in the first chapter of the Book of Genesis. The whole concept of extinction runs directly counter to the doctrine that God, having created the universe in a single event, made it perfect. However, while fossils present major difficulties for a conservative, literal reading of the Bible, a more liberal wing of Christianity has long since tried to come to terms with the scientific evidences of geological science.

The patterns of similarity and difference among living organisms has been a major focus of philosophical enquiry since Classical times. The 'Great Chain of Being' is a concept that goes back to Plato and Aristotle, and was developed by Descartes, Spinoza, and Leibniz. Here, everything in creation can be assigned a position relative to an ideal hierarchical pattern extending from nothingness at the base to God at the top. Man is next to God and the angels, the apes next to man, and so on. The lowliest bacteria (if they had known about bacteria) would have been at the base, just above minerals. In this hierarchy, each 'kind' is more complex and more perfect than, and in some way contingent upon, the one beneath. The 'chain' is static, the whole having been created by God, and it represents to us the perfect symmetry of his Creation. Any living organism can be given a place in the chain; potentially, any new discovery would readily drop into its place among the others.

The recognition of a vast world of fossils first supported and then challenged this view, as did the burgeoning scholarship and the first-hand knowledge of the living world produced by the explorations of the globe from the 16th century onwards. Soon there were too many kinds of organisms; at the least, instead of one chain, there must be many. The notion of linearity in the history of life was replaced by one of diversity – the Chain of Being became more like a tree of life. And, once it became obvious that there had to be several separate chains, it was necessary to posit the existence of organisms bridging them. As a medical student in Edinburgh (1826–8), Charles Darwin had his first venture into laboratory research with his mentor Robert Grant. Grant was a keen follower of the French zoologist Jean Baptiste de Lamarck's evolutionary ideas. Together Grant and Darwin studied 'zoo-phytes' – sea anemones – a group putatively sharing the features of both animals and plants.

A perfect Chain of Being would have no gaps; but while the growing fossil record closed up many gaps among groups, it opened up new ones and disclosed the existence of entirely new (extinct) groups. Extinction became a critical issue because it showed that the chain, or chains, could be broken. There were no obvious descendants of the newly discovered kinds of fossils with their wonderfully evocative names – giant reptiles likes mosasaurs, ichthyosaurs, and pterosaurs, and invertebrates like trilobites or graptolites. Those lines were dead ends. Perhaps most lines were. And many familiar living groups had extinct members, among the most dramatic of which were the mammoths and mastodons, unmistakably species of elephants but no longer living.

It is hard for us to understand the consternation of those generations of scholars and ordinary people who had to face the fact of extinction – that the living world does not represent the totality of 'creation' (however that word is meant) and the corollary that no life, ancient or modern, could be seen as complete and perfect; instead, life was changing rather than static. Many attempts were made to explain away extinction. The simplest was that we just have not looked everywhere in the world: somewhere there may still be living ichthyosaurs and trilobites. As justification for this view, John Ray wrote in 1693: 'Wolves and Bevers . . . were sometimes native of England (yet there remain) Plenty of them still in other Countries.' Thomas Jefferson, a true man of the Enlightenment, wrote descriptions of the mastodon fossils from Big Bone Lick in Kentucky. He thought that mastodons might still be living in the far West and hoped that the western Lewis and Clark expedition of 1804–6 would find them.

The concept of the Chain of Being depends on three premisses: plenitude (all possible versions of 'being' exist), continuity, and gradation. In the end, the enormous breadth of organismal diversity, in both space and time, simply overwhelmed theories of a static Chain of Being, however rationalized to extend to multiple creation events in space and time. The concepts of continuity and gradation made it logical for philosophers to ask whether organisms were not also related in the genetic sense, through a process of transmutation – evolution. In 1693, Leibniz (directly paraphrasing Robert Hooke's *Micrographia* of 1665) wrote that if extinction was a fact, it was 'worthy of belief . . . that even the species of animals have many times been transformed'. The Scottish philosopher David Hume, in his *Dialogues Concerning Natural Religion* (1779), asked whether it was more logical to assume that complex living creatures had their origins in simpler ones than via some miraculous creation by an infinitely powerful, but nonetheless unknowable, designing intelligence. During the Enlightenment, the Chain of Being became transformed into a 'Chain of Becoming' – a dynamic, temporal system involving some kind of historically contingent process.

In two early attempts to accommodate a sense of process without completely invalidating the static elements of the chain, the French philosophers and scientists Charles Bonnet (1720–93) and Jean Baptiste Robinet (1735–1820) changed the metaphor to one of a ladder of nature, the *scala naturae*. In the *scala naturae*, over time organisms naturally move up the ladder through a process of transformation. Thus today's mammals have moved up the ladder from a lower rung among the reptiles, and before that were fishes and worms. Today's worms have not yet moved on to become fishes or mammals. In Bonnet's system, various environmental conditions triggered the 'hatching' of a hierarchy of nascent germs set in place at the moment of Creation and lodged in the original 'souls' of organisms, causing them, step by step, to increase in perfection. In such cases, climbing the ladder involves some kind of literal or figurative unfolding or working out of a pre-ordained Divine plan. Such ideas remained true to the original teleological (end-directed) concept of the chain.

Erasmus Darwin (1731–1802, grandfather of Charles Darwin) and other late 18th-century thinkers saw the process more boldly, proposing that it was driven by continuous transmutation of species driven by changes in the 'generative systems' (that is, developmental genetics) of organisms and by the environment. Lamarck (1802) proposed a different scheme according to which there were many different, and always separate, lineages (not a tree, therefore, but a lawn). Each arose from the spontaneous generation of a very simple life form, the descendants of which then progressively climbed Bonnet's ladder according to a pre-ordained pattern of transmutation. In Lamarck's scheme, humans, being the most perfected organisms, belonged to the first-caused of these chains – and therefore the oldest and the one that has travelled the farthest towards the ideal goal of Godliness. Today's very simple kinds of living organisms have been created more recently and have only just started their journey. The overall pattern of increase in complexity seen in the fossil record is then explained as simply a matter of timing. In this scheme, no kind of organism truly becomes extinct, but currently may simply not be represented in one of the chains. One lineage or another of living reptiles will again produce ichthyosaurs, for example, or toothed birds. Charles Lyell, in his *Principles of Geology* (1831–3), and before being converted to

Darwinism, envisioned a sort of cycling in which the right environmental conditions would in the future produce extinct forms once again.

An old philosophical debate was then revivified. The place of humans in a static chain, created in a position next to God and the angels, was one thing; humans as the result of a state of material flux (due ultimately to the chance motions of atoms) was quite another. Not only did evolution remove the hand of God from our causation and reduce humans to the status of advanced apes, it opened up the whole question of the purpose and meaning of life. The discovery of a graded series of fossil species linking humans with the great apes would be a final irony for the Chain of Being. Linking humans in the other direction – to God – remained the province of religion.

'Awful changes. Man found only in a fossil state – reappearance of hthyosaurs.' Lyell's idea that life proceeds in cycles led Henry de la eche to produce this delightful cartoon in which a professorial hthyosaur is lecturing his contemporaries on the ancient history of umans

Fossils and change also have a political aspect. In the late 18th century, Liberté, Egalité, Fraternité, and, above all, progress were concepts that required not only a drastic revamping of contemporary social systems but depended on the raw material (in this case, humanity itself) having the flexibility and potential to achieve new goals, new stations. Older European oligarchies, on the other hand, depended on biblical authority for maintaining constancy and particularly the distinctions between hewers of wood and drawers of water, vessels of gold and silver, masters and servants.

As a model for social systems, theories of change that encompassed the whole organic world were naturally threatening to the establishment. Erasmus Darwin had the misfortune to develop his ideas of change during the French Revolution, when such ideas were unwelcome on British soil. But by the 1830s they had become unstoppable. Transmutation was further popularized in 1844 by anonymous publication of the quasi-Lamarckian *Vestiges of Creation*, authored by Robert Chambers, with its obvious reference to James Hutton's geology. Then came Charles Darwin, whose evolutionary theory explains the appearance of 'progress' in terms of process. During the voyage of the *Beagle* (1831–6), and influenced by reading Charles Lyell, Charles Darwin decided that he could make his mark as a geologist. He collected Pleistocene armadillos and tree sloths at Punta Alta, Argentina, in 1832 and realized that living species had replaced these older extinct forms over time. During his explorations of South America, he also saw that some living species replace each other in space – northern and southern pairs of species of rheas, for example.

Time

One of the principal lessons to be learned from geology and from fossils is that the earth itself is very old – some 4.5 billion years old (using the convention that a billion is a thousand millions) – and continually changing. What was once the sea floor is now mountains like the Alps or the great chalk cliffs at Dover, other ancient mountains have been ground down to sediments and redeposited in the sea, and so on in an unending cycle; the continents have moved, Europe and Africa having once been attached to the Americas. With all this came environmental changes that caused, for example, the tropical coal swamps of the Carboniferous (the products of which we mine in distinctly non-tropical places like Scotland and Pennsylvania) to wax and wane. All the time, life on earth was evolving. New groups of organisms steadily appear in the fossil record, replacing the old and sometimes opening whole new environments for colonization. In the Palaeozoic, plants, invertebrates, and vertebrates invaded the land for the first time. Insects took to the air, followed by the flying reptiles, the birds, and eventually the flying mammals. Different groups of organisms entered the deep seas, others found their way to mountain tops.

The record of the rocks, layer by layer, reveals, however imperfectly, the history of the earth and life upon it. The organisms that lived in ancient times were buried and preserved in sediments. It takes a long time for rocks to form and accumulate, layer-by-layer, on the earth's crust. Some rocks and the fossils encased in them are more than a billion years old, others are as new as the muds in which they are trapped.

The concept of geological time – both in the sense of an immense age for the earth, and the sense of processes acting on a scale that would be virtually undetectable in a human life span or even the record of human history (ecological time) – has a long history. Aristotle thought the earth was infinite in past extent and future duration. But the Judaeo-Christian tradition gives time a narrative form of beginning (Creation) and end (Judgement Day). The Book of Genesis records the Creation of the whole universe in an instant of time. On the other hand, philosophers like Descartes (1596–1650), as they pondered the possible origins of the earth and solar system, posited a more gradual, fiery beginning, and by so doing they made our origins, in principle, studyable in terms of modern concepts such as atoms, space, and motion.

There was a lot at stake politically as well as philosophically in such ideas. A philosophy strongly based on science inevitably threatened the authority of religion based on biblical authority. It is no coincidence that at the time of Descartes's death, just as scholars were worrying about extinction and getting serious about positing an ancient and progressive history for the earth and universe, James Ussher, Bishop of Armagh and Primate of All Ireland, provided priceless support for the authority of the Church against such heresies. He offered a 'smoking gun' for the literal truth of the biblical account of Creation. His calculations depended both on the genealogies recorded in the first five chapters of the Bible and considerations of the Julian and Hebrew calendars, the result being a definitive date of 23 October 4004 BC for the date of Creation. In fact, similar calculations had been attempted since the 1st century AD, and most authors had arrived at a date between 2000 and 4000 BC. Obviously 6,000 years was not enough for the sorts of events and processes that scientists were talking about; the idea of an ancient, changing earth must be wrong. Ussher's announcement of a definitive date for Creation was not only timely, it was well promoted: he managed to have it inserted into all editions of the King James Bible so that everyone would see it.

Despite Ussher, natural philosophers (as scientists were then called) continued to search for new kinds of truth. From the evidence of the rocks themselves, the Comte de Buffon (1707–88), among others, articulated a concept of uniformitarianism, according to which the processes by which the rocks were formed and changed, mountains eroded and then built up again, were the same as those processes observable (and, significantly, susceptible to rational study) today, the only difference being the time over which they have operated. There were no episodes of catastrophic intervention, miraculous or natural. The earth had arisen from a fiery ball of matter like the sun and progressively changed as it cooled.

<table>
<tr><th colspan="2">Aeon</th><th colspan="2">Era</th><th colspan="2">Period</th></tr>
<tr><td colspan="2" rowspan="13">Phanerozoic</td><td colspan="2"></td><td colspan="2">Quaternary</td></tr>
<tr><td rowspan="2">Cenozoic</td><td rowspan="2">Tertiary</td><td colspan="2">Neogene</td></tr>
<tr><td colspan="2">Palaeogene</td></tr>
<tr><td colspan="2" rowspan="3">Mesozoic</td><td colspan="2">Cretaceous</td></tr>
<tr><td colspan="2">Jurassic</td></tr>
<tr><td colspan="2">Triassic</td></tr>
<tr><td colspan="2" rowspan="7">Palaeozoic</td><td colspan="2">Permian</td></tr>
<tr><td rowspan="2">Carboniferous</td><td>Pennsylvanian</td></tr>
<tr><td>Mississippian</td></tr>
<tr><td colspan="2">Devonian</td></tr>
<tr><td colspan="2">Silurian</td></tr>
<tr><td colspan="2">Ordovician</td></tr>
<tr><td colspan="2">Cambrian</td></tr>
<tr><td rowspan="3">Precambrian</td><td>Proterozoic</td><td colspan="2"></td><td colspan="2"></td></tr>
<tr><td>Archaean</td><td colspan="2"></td><td colspan="2"></td></tr>
<tr><td>Hadean</td><td colspan="2"></td><td colspan="2"></td></tr>
</table>

4. Principal divisions of the geological timescale with dates and origins of major fossil groups

Epoch		Origins
Holocene		
	0.01	
Pleistocene		Homo
	1.8	
Pliocene		Apes and humans
	5.3	
Miocene		
	23.8	
Oligocene		
	33.7	Modern orders of animals and plants
Eocene		
	54.8	
Palaeocene		
	65.0	
UPPER		Flowering plants
LOWER		
	142	
UPPER		Reptiles flourish
MIDDLE		Birds
LOWER		
	205.7	
UPPER		Cone-bearing plants
MIDDLE		Mammals, dinosaurs
LOWER		
	248.2	
UPPER		Mammal-like reptiles
LOWER		
	290	
UPPER		Seed plant forests
	323	Reptiles
LOWER		
	354	
UPPER		Amphibians, insects, plants
MIDDLE		
LOWER		
	417	
UPPER		Fishes
LOWER		
	443	
UPPER		
MIDDLE		Colonization of land
LOWER		
	495	
UPPER		
MIDDLE		Most modern phyla
LOWER		
	545	
		Soft-bodied animals Algae
	2500	
		Bacteria
	4000	
	4560	

The Scottish philosopher-geologist James Hutton (1726–97) extended the principle of uniformitarianism and tried to calculate the age of the earth from measuring the processes of erosion and from the sedimentary record. The result was his classic *Theory of the Earth* (first outlined in his *Essay* of 1785), in which he followed Robert Hooke from a century before in seeing that the origin of new rocks from sedimentation and seismic activity was matched by weathering and erosion. Hutton recognized a dynamic process of recycling of the materials of the earth, resulting in a sort of immortality for the planet. But he could not discover a definitive age for the earth. In his most famous phrase, he concluded that geology revealed 'no vestige of a beginning, no prospect of an end'. He didn't mean there *had been* no beginning or that there *would be* no end. Rather, the constant churning of the earth's crust had obliterated the evidence.

Hutton's view fits rather well with modern ideas about the processes involved in 'plate tectonics', in which new material is forced out from the mantle at the places such as the submarine mid-Atlantic ridge where plates diverge, old continental plates are subducted at their edges, and whole continents are deformed where the plates collide. All the while erosion is eating away relentlessly at the continents, reducing all once again. It all happens very slowly; North America and Europe are currently moving apart from each other at a rate of some 3 to 5 centimetres per year (which on reflection is really quite fast). The Himalayas are being forced upwards by the colliding northward movement of the whole Indian plate at about the same rate.

Other scholars attempted to estimate the age of the earth from its rate of cooling. Descent into mines showed that the core was hotter than the surface. In 1863, the physicist William Thomson (Lord Kelvin) calculated, from the size of the earth and the rate of cooling, an age of 100 myr (millions of years) or less for the age of the earth. Now the scale of the argument had changed. Even a 100-myr age suited the arguments of 19th-century anti-evolutionists like Kelvin because it was not nearly long enough for the slow evolutionary processes that Darwin envisaged. Kelvin later revised his estimate to an even more hostile 40 myr. He did not know, however, that new heat was constantly being generated within the earth by nuclear processes, so his estimates were far too low. The modern estimate of 4.5 byr for the age of the earth is calculated from measurements of the rate of decay and proportions of radioactive isotopes in rocks.

Chapter 3
Fossils in the popular imagination

Throughout the 18th century, all educated people in Europe and the Americas were familiar with a broad range of fossils and many had a 'cabinet' of specimens. But it was still likely that their fossils would be classified as 'formed stones', a category that was neutral with respect to origins. Once it had been generally accepted that fossils were organic remains, however, they assumed an important role in popular culture as well as sober philosophical scholarship.

In the mid- to late 19th century, public fascination with ancient life was made possible by, and perhaps even helped precipitate, the popularity of inexpensive but well-illustrated publications for the mass market, such as Camille Flammarion's *Le Monde avant le Deluge*, published in Paris in 1886, and the development of the public museum. For two hundred years, fossils have provided the basis for a highly accessible kind of science. The phenomenon really burgeoned with the discovery of dinosaurs and a wide variety of other, often very large Mesozoic reptiles. My own first exposure to this popular literature was Arthur Conan Doyle's science fiction novel of 1910, *The Lost World*, although I cannot remember any ambition to become Professor Challenger.

Fossils have always attracted unusual and interesting people, not the least of them being Professor William Buckland at Oxford, the man whose family in the 1820s kept a bear in Christ Church deanery and who had a life-long ambition to eat his way through examples of the entire animal kingdom (he never found a good recipe for mole or house fly). Buckland's penchant for unusual pets helped solved the riddle of the fossil deposits of Kirkdale Cave in Yorkshire. Buckland realized the cave had been a den for hyenas; few other scholars of the day would first have hypothesized that the fossils in the deposit were not washed in there by the biblical Flood but represented a life assemblage. Not only did Buckland conclude that the many broken bones in the cave deposit had been cracked open by hyenas, he happened to have on hand a (more or less) tame hyena to test his theory and thereby became the world's first experimental palaeontologist.

Buckland was a popular and diverting lecturer on the subject of fossils and later in his career was elevated to Dean of Westminster Abbey. At the same time, only a hundred miles away but in an entirely different world, there lived someone who did even more to launch the popularity of fossils. Mary Anning (1799–1847) was – from economic necessity – one of the world's first full-time professional fossil collectors. It was she, apparently, who sold 'sea shells by the seashore'. As a young child she collected fossils on the beach to sell to the visiting gentry, as did other Lyme Regis residents. After her father Richard, an out of work carpenter, died, the 12-year-old spent most of her time on the beach and in the lower cliffs, searching for fossils.

Mary Anning offers a nice example of the timely convergence of people and places. Lyme Regis had become a popular coastal resort at the turn of the century and one of the attractions was the cliffs, from which waves and weather produced a variety of interesting fossils. Ammonites (or 'snakestones' – relatives of the living pearly *Nautilus*) were common, along with isolated vertebrae and what looked like crocodile teeth. The Blue Lias cliffs at Lyme Regis consist of layers of shale and limestone marl originally laid down (195–200 myr ago) in a shallow coastal sea. The fossils in the

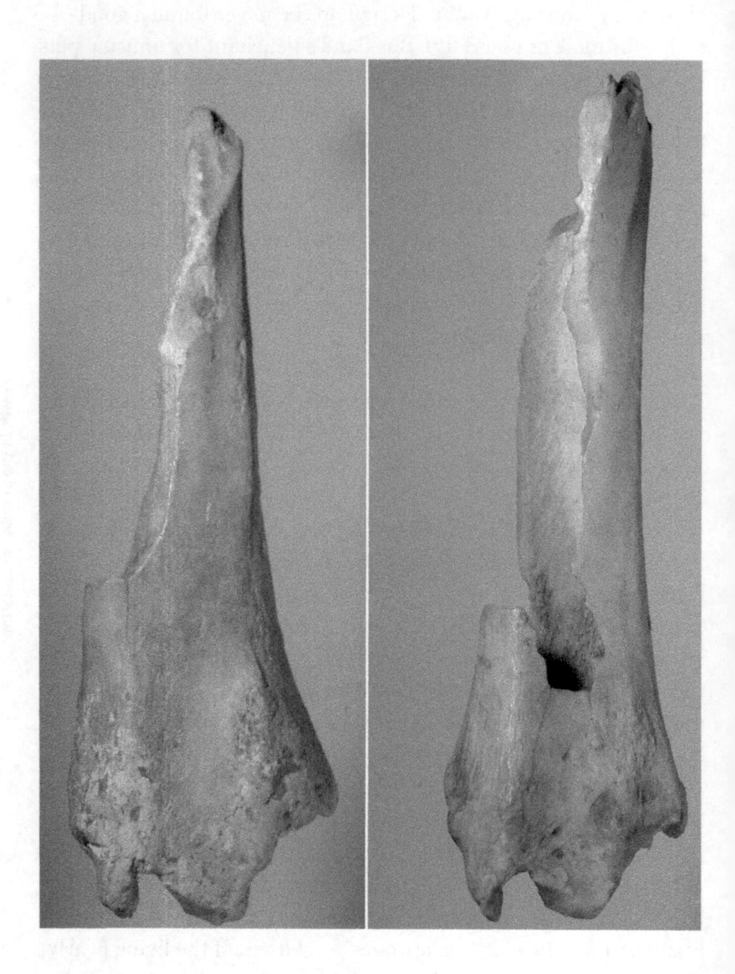

5. This fossil ox bone (left) from the Pleistocene Kirkdale Cave, first explored by Professor William Buckland in 1821, shows identical damage to that caused by a hyena on a modern ox bone (right)

limestones are preserved uncrushed and were especially sought after.

The cliffs there are soft. It may be that storms exposed a new fossil bed; perhaps it had been there all along and no one recognized it. In any case, between 1811 and 1812 Mary Anning and her brother excavated a large quite complete ichthyosaur (not the world's first, but the first to be described properly by scholars). She sold it for £23 to a local landowner. Ichthyosaurs, a sort of reptilian version of a toothed whale like a dolphin, turned out to be the source of all those 'crocodile' teeth. Later she discovered the first English flying reptiles (pterosaurs), the first plesiosaurs (a group forever fixed in our minds as including the Loch Ness monster), and a relative of the sharks that seemed to be a link to the skates and rays. Wealthy patrons vied to buy these new treasures, and palaeontologists in turn competed to be allowed to study and describe them, although Mary Anning always lived on the edge of poverty and a rigid class system kept her at the fringe of intellectual palaeontology.

Mary Anning

'. . . the extraordinary thing in this young woman is that she has made herself so thoroughly acquainted with the science that the moment she finds any bones she knows to what tribe they belong. . . . It is a wonderful instance of divine favour – that this poor, ignorant girl should be so blessed, for by reading and application she has arrived to that degree of knowledge as to be in the habit of writing and talking with professors and other clever men on the subject, and they all acknowledge that she understands more of the science than anyone else in this kingdom.'

Lady Sylvester, Diary, 1824

6. This sketch is widely taken to be of Mary Anning, with her hammer, on the beach at Lyme Regis

Not only do important discoveries tend to come in groups, so do the scientists. In the 1820s and 1830s, just as Mary Anning was finding amazing treasures at Lyme Regis, a whole new school of able palaeontological scholars emerged who would study them. In England, there were the Sussex doctor Gideon Mantell, Henry de la Beche (who later became the first director of the British Geological Survey), the Reverend William Conybeare, and of course Buckland. Meanwhile, in France, Baron Georges Cuvier, Professor of Anatomy at the Museum National d'Histoire Naturelle in Paris, and arguably the founder of vertebrate palaeontology as a professional discipline, dominated the scene with his encyclopaedic knowledge and magisterial opinions.

Buckland formally described the first dinosaur in 1824. In fact, Dr Robert Plot in 1667 described a large partial femur (now lost or strayed) from the Jurassic of Cornwell, Oxfordshire. Not knowing what it was, Plot had decided that it came from a biblical human 'giant'. Prompted by Cuvier, Buckland studied a new collection of material and showed that it came from a kind of giant reptile that he called *Megalosaurus* ('giant lizard'). He was quickly followed by Mantell, who found *Iguanodon* at a quarry in Tilgate Forest, Sussex. Both beasts were more than 30 feet long. *Megalosaurus* was evidently a fearsome predator, while, judging by its teeth, *Iguanodon* was a herbivore. In 1832 Mantell found another dinosaur which he called *Hylaeosaurus* ('forest lizard'). Then 10 years later the anatomist Richard Owen realized that there had existed a whole separate category of these creatures, not lizards at all and quite different from other kinds of reptiles, to which he gave the name Dinosauria (technically 'terrible lizard' – presumably on the grounds that 'saurian' could also mean 'reptile').

These extinct, extraordinary, but exceptionally real Mesozoic reptiles finally dispelled any possible notion that they, or any other fossils, were simply 'formed stones' – quirks of the rock mimicking living organisms. They were documented decades before Charles Darwin's *Origin of Species* and before any coherent theory or mechanism of evolution became widely acceptable, and some people still attempted to equate them with great mythical beasts such as the Behemoth of the Old Testament. Despite, or perhaps because of, their strangeness, ichthyosaurs, pterosaurs, plesiosaurs, and dinosaurs entered the popular imagination without difficulty. In fact, rather than interpreting them as subversive to biblical teachings, it was easy for people to see them instead as fitting neatly into an elaborated Chain of Being. The 1820s marked the peak of popularity for the movement of Natural Theology, in which the wonders of nature were studied as prime examples of the bounty and power of God. The Reverend Conybeare (the original describer of Mary Anning's plesiosaur) saw plesiosaurs as the link between ichthyosaurs and crocodiles and 'striking proof of the infinite richness of creative design'. Naturally, he dismissed as 'monstrous' the ideas of those who 'have most ridiculously imagined that the links (from species to species) . . . represent real transitions'.

But the whiff of gunpowder was in the air: the hint of powerful natural processes acting deep in the earth and deep in time could not be ignored. The evidence of the fossil record had already been one of the inspirations for Erasmus Darwin's theories of evolution. So sure was he of their importance that he even put some fossil shells on his newly minted family crest, along with the motto *E conchis omnia* ('everything from shells').

Dinosaurs

In 1801 Charles Willson Peale, a talented artist and even more talented showman, inventor of the modern natural history museum and the museum diorama, excavated two large and almost complete mastodons, and parts of a third, from Newburgh, New York. When put on public display in Philadelphia, Peale's mastodon was wildly popular and helped create the fascination for fossils among the public. But dinosaurs eventually took over centre stage. The very name contributed to their image, but their popularity also had a lot to do with the simple fact that they were so big and so 'different', while a mastodon was, after all, just another kind of elephant.

Right from the beginning, some workers on dinosaurs promoted their discoveries (and thereby themselves) in ways that workers with other fossils have not (or have failed to pull off). Richard Owen, for example, realized that dinosaurs were an obvious choice for display at Britain's Great Exhibition of 1851. Life-sized reconstructions of dinosaurs were created by the sculptor and master promoter Benjamin Waterhouse Hawkins for the great Crystal Palace in Hyde Park and these were later removed to a permanent site in south London. When a dinner party was famously held in Hawkins's half-built *Iguanodon* reconstruction in 1853, palaeontology moved a long way towards its modern media status.

Early dinosaur fever was even exploited by Charles Dickens, a man who was familiar with the power of cheaply available publications, and whose novels almost always appeared first in instalments in popular magazines. In *Household Words* in 1852, in the first paragraph of *Bleak House*, he wrote:

> Implacable November weather. As much mud in the streets as if the waters had but newly retired from the face of the earth, and it would not be wonderful to meet a *Megalosaurus*, forty feet long or so, waddling like an elephantine lizard up Holborn Hill.

Popular palaeontology never looked back, especially when the centre of action moved to the United States. The first American dinosaurs (isolated teeth from the Upper Cretaceous) were discovered in the Judith River Beds of Montana by the Hayden expedition of 1855–6. Then in 1858 the first partially articulated skeleton of any dinosaur was found in a New Jersey clay pit. Waterhouse Hawkins travelled to Philadelphia to mount Dr Joseph Leidy's new *Hadrosaurus* and then offered casts of it for sale to museums around the world. His mounted skeleton caused such a sensation that the Academy of Natural Sciences instituted museum admission charges to limit attendance. A copy was exhibited at the American Centennial Exposition in 1876 and later at the Smithsonian.

But the great surge in discovery came with the opening of the American West. Once again, several scholars converged on the same subject. The gentlemanly Leidy (once described as the last man who knew everything) had been supplied with fossils from the West by a number of explorer-collectors, including Ferdinand Hayden. But he was soon eclipsed by a group of well-funded, thrusting scholar-adventurers who, even where there was room for all, competed bitterly for the fossils of the West. In 30 years the arch-rival collectors and scholars Othniel Charles Marsh at Yale University and Edward Drinker Cope in Philadelphia collected some 120 different dinosaurs alone from the badlands of the American West. Together they collected countless other kinds of fossils too (although they tended to look for the dramatic material and to overlook other, equally interesting things such as primitive mammal teeth).

The story of Cope and Marsh is one of the great sagas of science, at turns funny, reprehensible, and tragic. But there was no doubting their determination. In 1875, not only was Marsh in the Black Hills of South Dakota negotiating with the Sioux for permission to collect, he quickly became the advocate in Washington for Red Cloud against the neglect of the US Indian Agency. In 1876, just a few weeks after the Battle of the Little Big Horn and the defeat of Custer, Cope was collecting in Montana, reckoning that 'since every able-bodied Sioux would be with the braves under Sitting Bull . . . there would be no danger for us'. Exploration and adventure were central to the pursuit of the scientific riches of the West, and those same landscapes were captured for the public in the paintings and prints of artists such as Alfred Bierstadt, Thomas Moran, Karl Bodmer, and George Catlin.

At the turn of the 20th century, Western American fossil explorations extended to the contiguous, and no less dramatic, badlands of Canada. The charismatic Charles Sternberg, who worked for both Cope and Marsh and then for the Canadian government, collected extensively in the Red Deer Valley of Alberta. The fabulous fossil resources of this region are now displayed at the Tyrrell Museum in Drumheller, Alberta.

In terms of public interest, dinosaurs took over from pterosaurs and ichthyosaurs with the spectacular Belgian discovery of a mass grave of *Iguanodon* in 1878. In 1897, a sensational, lavishly illustrated article, 'Gigantic Saurians of the Reptilian Age', in *American Century*, taken up in the tabloid *New York Journal* and *New York World* in the following year, finally established dinosaurs as a phenomenon of popular culture.

In 1895, the American Museum of Natural History purchased the private collections of the bankrupt Cope from under the noses of the Philadelphia Academy, and director Henry Fairfield Osborn (himself a palaeontologist) made the dinosaur one of its showcase images and attractions. In 1902, an expedition led by Barnum Brown bagged the first *Tyrannosaurus rex*. Then, in the 1920s and 1930s, the Museum opened things up much further by sending a series of expeditions (not forgetting the movie cameras), led by the dashing Roy Chapman Andrews (the prototype 'Indiana Jones') in his highly polished riding boots, to the Gobi Desert. Their original goal had been to search for early humans; instead, they made startling discoveries of horned dinosaurs and nests with eggs still inside.

Not to be outdone, the Carnegie Museum in Pittsburgh, with the financial backing of its eponymous founder, launched its own major research efforts, as did all the other major museums, buying from collectors if they did not mount their own expeditions. Today, 75 years on, the search for yet more important dinosaurs (and other fossils) has been successfully extended to the whole world, from the Arctic to Argentina, China to Greenland, Australia to Africa. And interestingly enough, the rise in popularity of palaeontology has brought a resurgence of dinosaur discoveries in Britain, particularly in the Isle of Wight. Wherever anyone collects, dinosaurs are most likely to attract the headlines.

The last romantics

Few sciences have been as successful as palaeontology in remaining serious and yet broadly accessible at the same time. A great deal of its popularity may come from the image of the palaeontologist as explorer and therefore a 'player' in a world that seems glamorous and exciting. The prevailing popular image of the palaeontologist is of the rugged individualist. The noble explorer pits himself against the wilderness and brings back fabulous things.

A great deal of this image is true and partly stems from the fact that, at the end of the second half of the 19th century, dinosaurs, and with them a great deal of other palaeontology, entered the myth of the American West. No longer were important discoveries made by gentlemen in suits and ties (perhaps having removed their jackets) directing a couple of workmen in a small quarry in England or New Jersey. Instead, fossil collecting had become 'prospecting'. A man with a horse and a pick – and of course a rifle – could venture out West and, like his gold-seeker cousins, bring back untold wealth from the rocks. That is an image that carries a great deal more weight than the reality of the man or woman in a white lab coat, doggedly extracting tiny details from trays of museum specimens and preoccupied more with complex statistical methods and the chemistry of sediments than with campfires in the badlands. No matter that the vast proportion of palaeontology was being conducted in less than glamorous conditions and concerned most undramatic organisms like graptolites and brachiopods. Long after it had become largely a laboratory science, palaeontology was depicted as a matter of rugged individualism and enterprise, richly rewarded.

Palaeontological tradition after the turn of the 20th century was therefore built under a dual rubric: the formal laboratory science typified by the development of the research universities and the dying embers of the Romantic Movement. Thereafter, every summer, professors from the great scholarly institutions would throw off their dark suits and ties for the casual shirts, jeans, and boots of the prospector. And each autumn they brought back their fossils to the laboratory. Each was an Othniel Charles Marsh, Roy Chapman Andrews, or (latterly) Indiana Jones. For the public and for many palaeontologists, professional or amateur, fossils continue to represent this happy fusion between 19th-century romanticism and the cold, hard clarity of contemporary science.

Chapter 4
Some things we know, some things we don't

Palaeontology is a dynamic and exciting science because we still know the 'fossil record' imperfectly. Each fossil is only a partial representation of the original living organism; fossils collected together incompletely represent their original communities or lineages. Our ideas about fossils are incomplete too; we still have much to learn from them about the history of the earth and about the great central theory of all biology, the theory of evolution. The fossil record is full of extinct organisms far beyond the extreme imaginings of science fiction. And almost as full of gaps. Some species are known from literally thousands of specimens, some from only a handful. Major discoveries remain to be made, with the potential to open up new lines of thought or to refute old ideas. Each new fossil is the equivalent of a pixel in a very large image – a developing image of ancient, once-living worlds. Fossils are our only way of recovering those worlds.

A common misconception about the fossil record is that it documents a smooth process of unfolding, evolving, biological diversity. For some groups, such as mammals and birds, and for some major transitions such as the origin of tetrapod vertebrates (amphibians, reptiles, birds, mammals – throughout this book I use the terms in their old-fashioned, popular sense) from fishes – even for the origin of humans – recent discoveries have significantly advanced our understanding and filled in many gaps.

But the fossil record is also replete with groups of organisms that appeared quite suddenly, flourished, and then disappeared again. While the larger group (the phylum or class) to which they belong may be known, their evolutionary parents cannot be discerned in the record as we currently know it and they appear to have left no descendants. Examples of these apparently rootless groups include some of the most successful of ancient life forms, such as the strange Palaeozoic fishes called acanthodians (something like extra-spiny sticklebacks) and heavily armoured fishes called placoderms, peculiar but extremely abundant invertebrates such as graptolites and trilobites, and dozens of less familiar creatures. Here there is much work to be done.

Reading the rocks

After Robert Hooke and Steno had established that fossils must be the remains of once-living organisms, fossil-collecting gradually became a systematic science and assumed huge importance in practical geology. Steno's three great principles of 'superposition', 'horizontality', and 'lateral continuity' made it possible to open the geological record like a book: fossils made it possible to read and number the pages.

From the 18th century onwards, as the inner structure of the earth became more open to view as a result of deeper mining and the building of roads, canals, and railways, it quickly became obvious that certain kinds of rocks were associated with particular fossils. The coal beds of the Carboniferous and Eocene, for example, contain entirely different sets of fossils. The stratigraphic column was eventually seen to consist of superimposed layer upon layer of rocks containing their own signature fossils, the most useful of which are microfossils like forameniferans (microscopic one-celled planktonic organisms, modern species of which still live in the ocean, and whose skeletons are deposited by the billions on the sea floor).

These signatures can signal differences in both time and space. For example, two fossil beds might well be of the same age, but if the original ecologies and depositional settings were different – for example, if one was an ancient lake bed and the other a coral reef – they will contain totally different fossils. However, the reverse is also true, and exceptionally useful. If we have two outcrops of rock, miles apart, that have the same fossil signature, we can be sure that they are the same age and were originally laid down in similar conditions.

From the consistency of the 'fingerprints' of fossil distribution in sedimentary strata, the English surveyor and canal-builder William Smith (1769–1839) discovered that he could trace the same Jurassic rocks that outcropped in Somerset where he had been involved with canal-building in a diagonal line north-east, right to the North Sea coast. Criss-crossing the country collecting fossil and rock samples, he created the first maps of the surface geology of Britain.

Steno's laws

Superposition: At the time that any Body was formed, there was another Body under the same Bed [that] had then already obtained a solid Consistency.

Horizontality: 'Tis certain that, when any Bed was formed, its inferior surface, and that of its sides, did answer to the surfaces of the inferior Body; but the superior surface was, as far as was possible, parallel to the Horizon.

Lateral continuity: At the time that any Bed was formed, it was *either* at the sides environed by another solid Body, *or* it did cover the whole Globe of the Earth.

Steno, *De Solido Intra Solidum Naturaliter Contento Dissertationis Prodromus* (1669; English tr., Hans Oldenbergh, 1671)

Furthermore, in many places Smith could also map the subsurface geology. Such maps have immense value. They predict which kinds of rocks will be under the surface even where it is covered by vegetation, they inform the reader what those rocks will contain – iron ore, coal, building stones, conditions where railways can be built, where canals should not. For all this, the fossils were crucial.

The entire geological timescale as we know it, from Precambrian to Holocene, was originally based on distinctions among these fossil signatures – whether it be the major divisions like Palaeozoic or in the finest details. There is, of course, a problem of circularity if we use the rocks to date the fossils and the fossils to date the rocks. And, if it were possible that any species could have arisen twice, independently, then we will have been fooled. Luckily, some rocks can now be assigned dates by use of radiometric methods, thus providing both an absolute age and an independent calibration of the relative ages provided by the faunal and floral record.

With the stratigraphic column recognized as a unique time-sequence, the overall patterns in the fossil record became clear. The history of life on earth was revealed as 'progressive', at least in the sense that it started from a beginning in simple (so-called 'lower') organisms and proceeded via ever-increasing diversity and complexity. Perhaps the most telling aspect of this new fossil record was that the overwhelming majority of its denizens were not to be found living on earth today. The deeper one went into the rock strata, the more extinct forms emerged – not just extinct species but whole major groups (trilobites, for example). All these discoveries not only provided new facts to be assimilated by scientists, they caused a huge stirring of the popular imagination. Whole other worlds had waxed and waned before our own.

From the palaeontological viewpoint, living faunas and floras are revealed as the latest manifestations of a story of organic change that has proceeded for at least 3.5 billion years. Once life had evolved, at every age in the history of the earth, a different signature of life existed; in the familiar Geological Time Scale, with its dates calibrated by radio-isotopic methods, their fossils now define the constituent units, large and small.

The general outlines of the changing course of life on earth over the immediately past 545 million years of the Phanerozoic ('evident life') Aeon are well known. If we trace *backwards* in time from the present day, most of the groups of organisms that we are familiar with today can be shown to have arisen in the Cretaceous and diversified within the last 66 million years - the Cenozoic Era (originally spelt Caenozoic or Cainozoic, and meaning 'recent life'). This was the time of broad diversifications among early representatives of the mammals, birds, fishes, insects, grasses, and of flowering trees and plants co-evolving with the insects that pollinated them.

If we look further back into the Mesozoic Era ('middle life'), we see a very different world, the denizens of which - including entire groups like ichthyosaurs, plesiosaurs, flying reptiles, and ammonites - are extinct. In this 'Age of Reptiles', the forest trees were principally conifers and cycads. And, of course, there were the dinosaurs, one branch of which may survive today in the form of birds. Some familiar kinds of animals, like crocodiles and turtles, were well represented, and modern groups of fishes were beginning to diversify. Various kinds of primitive mammals existed from the Late Triassic onwards.

Looking further back still, into the Carboniferous Period of the Late Palaeozoic Era ('ancient life'), we enter a far more archaic world. On land there were tropical forests of huge trees, including forms related to modern horsetails and lycopods, whose remains gave us coal, and swamps that were home to giant insects and bizarre amphibian tetrapods. In the seas vast limestone deposits were being laid down. A yet more primitive world existed in the preceding Devonian times, often described as the 'Age of Fishes'. This, beyond all others, is the world that I would wish to visit in a time machine, a largely barren landscape where our earliest four-legged amphibian ancestors eked out a precarious existence mostly at the marshy edges of the continents and around inland lakes and swamps. This was also when the first vascular plants and wingless insects ventured on land, away from rivers and seas dominated by a variety of strange, heavily armoured fishes, including predators 20 feet long.

Further back still, the Cambrian, Ordovician, and Silurian have often been called collectively the 'Age of Invertebrates', being dominated by trilobites, brachiopods, the odd graptolites, and the even stranger creatures called conodonts (usually found represented by isolated teeth) which may have been chordates, suggesting that the ancestors of all the vertebrates had already evolved in the Cambrian.

None of this happened in a geological-geographical-ecological vacuum: right from the beginning, predator and prey, consumer and food, have had to engage in the equivalent of an arms race. New defences against being eaten were followed by new ways of defeating them. For example, one of the most striking features of Tertiary life is the evolution of hard silica crystals in grasses, and the consequent evolution of kinds of teeth in ungulates (cattle, deer, horses, for example) and rodents to deal with those protective abrasives.

All these changes among living organisms were accompanied by, and in substantial part driven by, physical changes operating on a truly global scale. The earth's crust consists of a number of 'tectonic plates' that move around over the semi-fluid mantle beneath. In the Proterozoic all the continental plates were probably combined into one super-continent. By the Ordovician, there was one large continent – Gondwana – consisting principally of a pushed-together Africa, South America and Australia sitting somewhere in the great southern ocean, while Asia, Europe, and North America were more or less isolated in the northern hemisphere and largely covered with seas. By the Devonian, Western Europe and eastern North America had joined as a large northern land mass called Laurentia and, by the end of the Devonian, the continents were coming together into one great super-continent (called Pangaea). Then, in the mid-Mesozoic, they started to break up again, the Atlantic opened up, and the continents edged towards their present positions.

The implications of these global movements are still being analysed. But it is clear that plate tectonics is a major set of factors in driving organic evolution by producing the environmental context for a great deal of evolutionary change. Mountain-building, opening and closing ocean basins, creation and loss of shallow continental seas, and sea-level changes affected (and continue to affect) climate through the balance of tropical, temperate, and polar land environments, major ocean current systems (for example, there could be no Gulf Stream before the Atlantic Ocean opened), and atmospheric circulation (uplift of the Himalayas may well have been the origin of monsoons).

Undoubtedly the fossil record will continue to change; many new kinds of creatures will be found and the records of known ones will be pushed further back into deep time. But it seems unlikely that the general picture of change in life on earth during the relatively recent times of the Phanerozoic (the last 545 myr) will be falsified. Many challenges remain, however, when it comes to earlier times still.

In the beginning

The biggest gaps in the fossil record are right at the beginning. No rocks survive in the earth's crust that are older than about 3.9 byr, all earlier rocks having been recycled by earth processes. The period between the earth forming some 4.5 byr ago and the beginning of the Cambrian, 545 myr ago, is defined as the Precambrian Aeon, divided into three Eras (first the Hadean; then the Archaean, beginning 4 byr ago; and most recently, 2.5 byr ago, the Proterozoic, meaning 'first life').

A small amount of evidence, mostly still controversial, records the presence of bacteria and perhaps other microbial life in Archaean rocks from Australia and South Africa dated at 3.5 byr ago. The principal kinds of bacteria were cyanobacteria: the name refers to the blue-green colour, not the production of cynanide. Cyanobacteria are still abundant on earth today. There is, however, a gap of at least a billion years between the formation of the earth and these first signs of living organisms. At some point in that interval, life arose on earth in the form of relatively simple self-replicating molecules and proceeded to the formation of something like modern viruses and bacteria.

Between 3.5 byr ago and the beginning of the Cambrian, there arose a new kind of organization in the form of single-celled algae. These occurred in mound-shaped structures called stromatolites that were composed of layers of cyanobacteria, algae, and sediment. Similar colonies are found today in tropical, shallow water lagoons such as at Shark Bay in Western Australia, and it may even be that the same species are involved.

The presence of cyanobacteria and algae in the Proterozoic means that, for the first time, photosynthesis had been achieved. Photosynthesis is the process by which all plants today tap the energy from the sun to power the synthesis of sugars and other carbohydrates from the simple molecules carbon dioxide and water. These first 'primary producer' plant-like forms, or autotrophs ('self-feeding'), could capture the almost limitless energy from the sun and store it. Eventually they provided thereby the base for a food chain, following the evolution of a new range of organisms, heterotrophs ('other-feeding'), that we call animals; they evolved to get their energy by eating the autotrophs. With this, the dog-eat-dog nature of complex ecosystems had been set in train.

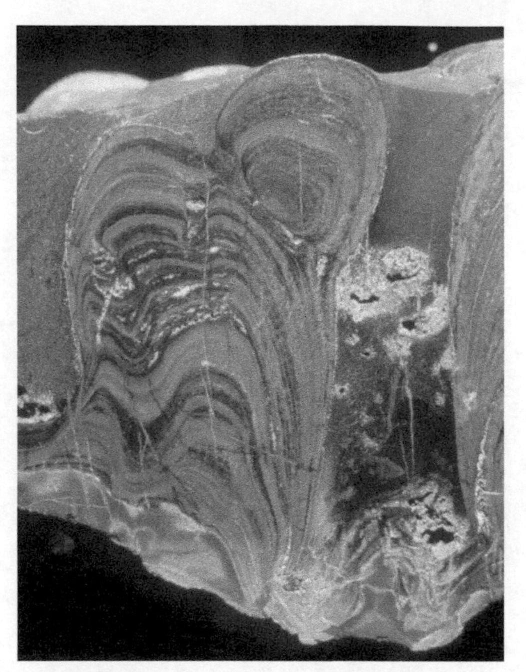

7. Stromatolites consist of layers of algae and sediment, as shown in this cross-section of a Precambrian form

One important aspect of photosynthesis is that plants release oxygen. An essential feature of most heterotrophs – and all the different kinds of animals we know today – is that they need oxygen in order to break down complex molecules to release their energy (essentially to reverse the process of photosynthesis). The waste oxygen from plants therefore supports animal life. From 2.5 to 2.0 byr ago there seem to have been relatively low concentrations of free oxygen in the atmosphere. This is indicated by the deposition, worldwide, of 'banded iron' deposits consisting of alternating iron mineral (magnetite) and chert (silicon dioxide). The phenomenon was evidently the result of a cyclic process involving dissolved iron (derived from weathering of sediments) and oxygen-producing bacteria in the early oceans.

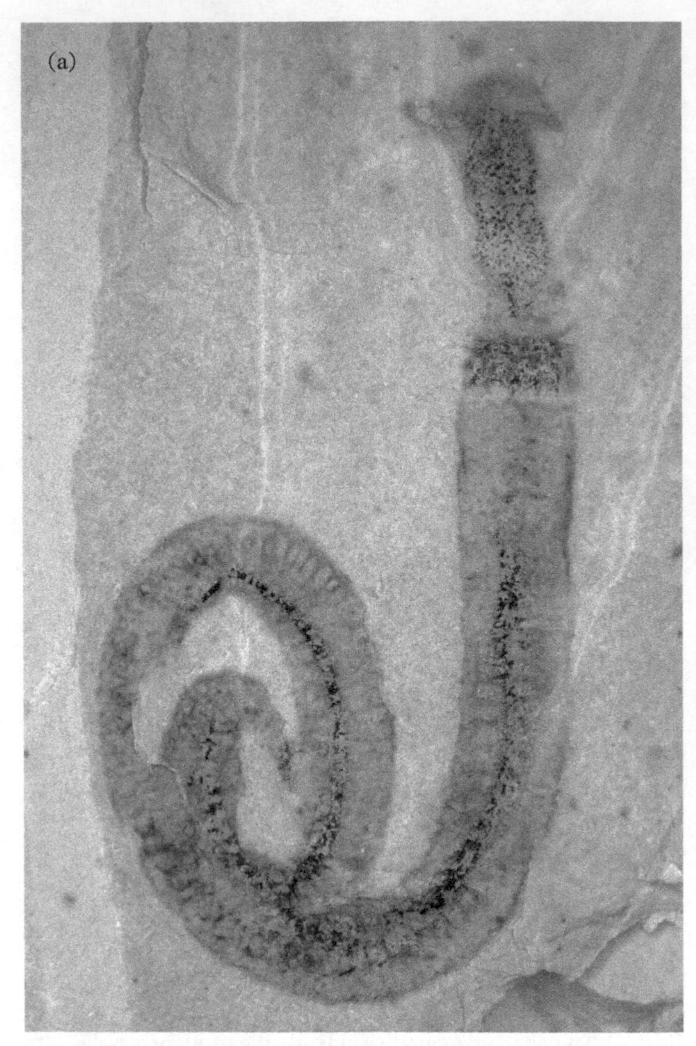

8. Early Cambrian fossils: (a) *Circocosmia*, a kind of worm from Chengjiang, China and (b) *Sanctacaris*, a spider relative from the Burgess Shale, Canada

The iron combined with oxygen and was precipitated as layers of magnetite a few millimetres thick, arguably the process being halted when oxygen supplies were depleted. By about 1 byr ago, a sharp reduction in banded iron deposition shows that significantly higher levels of atmospheric oxygen had built up.

Seemingly paradoxically, oxygen, which we animals regard as essential for life, is actually a chemical poison unless tamed – and exploited. The new high levels of atmospheric oxygen killed off most of the archaic older kinds of bacteria and algae. This was accompanied by a new explosion of evolution in the form of different kinds of much more complex, multicellular organisms using oxygen in the new way. So far, most life had been probably concentrated in shallow seas and coastal margins: but around this time we start to find a new kind of small single-celled algae called acritarchs that lived in the open ocean, which meant that another major global ecosystem had opened up. This in turn led to yet more oxygen production.

Some of our most intriguing information about the diversification of life in the early Cambrian comes from a small number of sites (called 'Lagerstatten', or 'mother lodes') where soft-bodied forms have been preserved. Of these, the Middle Cambrian Burgess Shale in British Columbia (discovered in 1909), and the slightly older (525 myr) Chengjiang site in Yunnan Province, China (discovered in 1984), are the most important. Present are several kinds of multicellular algae and at least 12 different major groups of animals, including familiar phyla such as the Porifera (sponges), Coelenterata (jellyfish), Mollusca (molluscs), Annelida (worms), Arthropoda (relatives of crustaceans and insects), Echinodermata (relatives of starfish and sea-urchins), even a diversity of early Chordata (our own relatives) sitting somewhere near the base of the tree of vertebrate life. Many of these fossils are difficult (and therefore controversial) to interpret, and some have been given appropriate names like *Anomalocaris* and *Hallucigenia*. Most interestingly, not only had the great diversity of living forms that we know today already largely been set in place, many lineages (and perhaps even whole phyla) had apparently arisen, flourished, and then become extinct in Precambrian and Cambrian times. The period between 600 and 500 myr ago had been one of intense experimentation in new ways of being an animal or plant and a corresponding culling of many groups.

Earlier geologists chose to fix the beginning of the Cambrian at the point at which they found the first apparent burst of life on earth (with the Precambrian having erroneously been defined as having been without life). However, linking the sparse record of the Proterozoic with the richer Phanerozoic are a small number of localities where an astounding fossil record has been found – a record so rich and full that we can only guess at the sequence of evolutionary events that produced it. The Ediacaran faunas (named after the most important locality in the Ediacara Hills of South Australia), ranging from about 565 myr to 540 myr ago, are very difficult – and contentious – to interpret. Some palaeontologists think that they include impressions of jellyfish, molluscs, worms, proto-arthropods, echinoderms, and possible sponges. Others see them as a range of forms whose affinities are still not clear but that are basically unrelated to modern lineages. One interesting element of the faunas is the absence of obvious carnivores; all the known Ediacaran animals appear to be bottom-dwelling suspension and detritus feeders. Most of them had died out by the beginning of the Cambrian.

9. The frond-like Ediacaran *Charnodiscus* fossils from the Mistaken Point Formation in Newfoundland were filter and/or detritus feeders

Evidently, a hard skeleton had not evolved in any Proterozoic lineage of multicellular organisms (which is palaeontologists' shorthand for 'we haven't found any yet'). It seems that at some point before 545 myr ago, an environmental trigger had promoted the development of hard tissues – usually minerals based on calcium or silicon – and this allowed the survival of a better fossil record. There is evidence of at least one major global-scale environmental event in the Proterozoic that may have been involved in promoting hard tissue formation. Between about 700 and 600 myr ago, there seem to have been one or more major episodes of global cooling. These resulted not merely in massive glaciations of landforms but possibly also a complete deep freeze such that even the seas at the equator iced over. Under such conditions, life could have survived only in places like hot springs and deep-sea volcanic vents. Whether or not the cold extended to a full 'snowball earth', as some hypothesize, or just a 'slushball', the effects on the fledgling faunas and floras must have been significant. Warming after the freeze was probably driven by plate tectonic upheavals, producing earthquakes and volcanoes that released large volumes of greenhouse gases. This resulted in wholesale weathering of rocks that released a huge amount of calcium carbonate and phosphate into the oceans – calcium with which to make skeletons.

Chapter 5
Against the odds

I collected my first fossils as a graduate student, but I had looked for them long before. In my childhood, our large draughty house was heated by coal. I knew that people found fossils in coal, so I would retire to the coal store and pound away with the large hammer kept there, hoping to find a lovely fern frond or an amphibian tooth. (My getting covered with coal dust was tolerated because someone had to break up the large lumps anyway.) Wherever people had found Carboniferous fossils, however, it was not in the places where our fuel supply came from. But still I hammered away, driven by the lure of buried treasure. Only later did I learn that the best fossils are found in the shale layers between the coal seams and that, in any case, I would have had to split many tonnes of rock to find even a single recognizable macrofossil. Even in the Coal Measures, where the coal itself is (largely unrecognizable) fossil plant material, the odds are against you.

Out in the field, there is a magic moment when a fossil is first spotted. Typically, you have been searching for days without success and then you look down and see a gleaming corner of shell or bone partially exposed in the ground. Perhaps a series of fragments slipping down the rock slope has led you up to where the main part sits. Now you must get it out. Without risking any damage, you carefully work around it with a chisel or a sharp knife. Gradually you find the limits of the thing – is it just a broken piece, or does the whole fossil lie there in the rock? Moment by moment, sometimes hour by hour, or even day by day for a really big object, the fossil shows itself. Most often it is nothing special, but sometimes it is a skull, a superb coiled ammonite, a limb bone, a group of clam shells, a shark tooth, the leaf of an ancient plant. Once it has been exposed, it may need to be treated with a hardener – a lacquer that can easily be removed in the lab. Something small and hard like a trilobite or a shark's tooth might be slipped into a plastic bag, protected with tissue paper; a roll of toilet paper is invaluable for this. If the fossil is any size at all, it will be best to make a jacket for it out of strips of cloth soaked in plaster of Paris – just like an old-fashioned cast for a broken leg. First, the top is covered and then comes the point when it has to be turned over. You slide a knife underneath and gently lift . . . and out it comes, without crumbling to pieces. Finally you have it, ready to take back to the lab, neatly marked with its own number and listed in the field book. And then, almost greedily, you look for more.

There is no way of knowing how many fossils have been collected over the years. The fossils still remaining in the rocks are beyond counting, and they themselves represent just a tiny faction of the diversity of life that has ever lived on earth. Each fossil represents only a small portion of the once-living organism – usually only those parts of it that were tough and resistant to decay. While popular attention is usually focused on the fossil bones of vertebrates like dinosaurs and humans, the most common kinds of fossils are the remains of the skeletons of marine invertebrates and the calcareous or silicaceous shells of microscopic plankton.

What is a fossil?

The best way to know what fossils *are* is to examine how they are *formed*, the study of which has become a major branch of palaeontological science.

The chances of any organism being preserved as a fossil are remote:

many millions to one against. Once any organism dies, its long, sun-powered struggle against entropy is over and decay sets in. There is only a tiny chance of it surviving in the ground or under water before being consumed by large scavengers and small, and the inexorable attentions of bacteria and fungi. In nature, many more processes act to remove dead organisms from the environment than to preserve them. On the whole this is a good thing. For example, there are some 700,000 roe deer living in Britain. If each lives for 15 years, then on average nearly 50,000 deer die every year. If this were not the case, there would be an epidemic of live deer, but unless the carcasses were quickly broken down, we would soon be buried in dead deer.

For a fossil to be formed, whatever survives the early processes of decay has then to become incorporated within sediment that eventually becomes rock. At the same time, the remains will undergo the chemical and physical changes (diagenesis) that will transform them into 'rock' also. The most likely parts of any organism to be preserved as a fossil are the obvious hard parts – the mineralized shells of invertebrates like ammonites and molluscs, the bodies of corals and sponges, the bones and teeth of vertebrate skeletons, the chitinous exoskeletons of arthropods such as trilobites, crustaceans, and insects, and the woody tissues of plants. Tree trunks are quite frequently preserved. But soft-bodied organisms such as worms and jellyfish are far less likely to be preserved as fossils. Numerically the most common fossils are surely the microscopic shells of marine planktonic organisms and the tough-walled spores and pollen of plants.

Very little of the original soft tissue of any organism is preserved in a fossil. For example, sharks have very hard, mineralized teeth, and sharks' teeth are common as fossils, but remains of the rest of the body are rare. However, under certain conditions, whole communities of soft-bodied forms may be preserved. This happened in the case of the Cambrian-age Chengjiang and Burgess Shale Lagerstatten. Track-ways made by animals that have crawled over, or burrowed into, some ancient mud turn up surprisingly often. In relatively modern shells, sometimes colours are preserved; sometimes the original biochemistry persists in the form of amino acids, and even segments of DNA in very young fossils. Despite popular fiction, however, it is not possible to recreate extinct creatures by genetic engineering of fossil genes.

The environment is critical too: whether an organism becomes a fossil after death depends both on the nature of the original organism and in part on the circumstances and environment of its life and death. Sessile organisms (like corals and molluscs) are more likely to be preserved than free-swimming forms in the same community (fishes, cephalopods, crustaceans). Organisms living in 'high-energy' environments such as fast mountain streams, or where there is heavy wave action, are more likely to be destroyed physically after death than those living at the bottom of the sea. Animals and plants dying in woodland environments fall into a rich leaf litter full of organisms that live by decomposing organic material and are likely to be turned into compost quickly. In the soil, not only is there biological breakdown of tissues, numerous soil organisms burrow around disturbing everything and breaking up fragile tissues. Organisms living in all of these environments will be far less likely to become covered by layers of sediments than those that die in or near river deltas or shallow marine lagoons. Insects and other fragile terrestrial creatures are less likely to be preserved than shellfish and heavy-boned vertebrates. Animals like worms and jellyfish and plants like algae and mosses that lack any kind of solid skeleton (mineral, protein, or wood) will be broken down fastest of all. The study of all these processes and conditions in fossil formation is termed 'taphonomy'.

Before burial

Earth to earth, ashes to ashes, dust to dust – from the moment that the organism dies, a race begins between the processes of decay and dissolution and those that tend to preserve whatever survives. First, the remains must survive those pre-burial events that reduce the remains of most organisms almost to nothing in just a few weeks. Scavengers may eat the carcass, insects may lay their eggs there and the larvae will hatch out further to consume the rotting remains. The majority of decay is caused by bacteria, both from within and without the corpse. If the remains stay in a moist environment or are washed into a body of water – and of course many organisms live and die in water to start with – saprophytic (tissue-destroying) fungi will also be a major cause of decay. Organisms that are deposited in terrestrial soils, already rich in decayed and decaying organic matter and full of organisms that thrive there, are particularly vulnerable (compost to compost).

Breakdown of tissues will be accelerated by warmth. The various bits may become separated as decay loosens hinges and joints. Scavengers may tear carcasses apart and carry the parts away. Water currents then distribute the pieces differentially downstream or along a beach, the smaller parts being carried further. Plant leaves and pollen, which might otherwise be highly indicative of particular ecological conditions, may be distributed over great distances by the wind, making the task of the palaeo-ecologist particularly difficult.

Some fossils may actually record elements of these pre-burial processes. Tooth marks of predators and scavengers are sometimes preserved on bones and shells. Fishes are often preserved with their abdominal cavities burst open due to gases produced by bacteria. The sorting and/or orientation of fossils on the bedding plane may indicate the current direction and force.

Various pre-burial factors may prevent or delay all this dissolution, destruction, and decay. In general, our best chance of recovering information about a natural community in life may come through some decidedly un-natural aspect of its death. Organisms may die in conditions that are toxic to scavengers and bacteria. The simplest such case might be mummification through rapid aerial drying. An insect captured in a dollop of plant resin that becomes amber is essentially pickled. Peat bogs, where there are high levels of humic and tannic acids that retard or prevent bacterial action, tend to preserve remains very well and are basically the source of coal formations. There are instances of humans being superbly preserved in modern peat bogs. The earlier and longer that any such factors apply, the greater the chance that organisms will be preserved intact and in many fossil beds individual specimens are still *in situ*.

One of the most dramatic examples of such special conditions is the ancient oil seeps preserved at the Rancho la Brea tar pits of Southern California. Starting around 40,000 years ago, pools of tar seeped to the surface. Whenever they were covered with pools of rainwater they became an attraction to wandering animals like the native horses (not yet extinct in North America), camels, mammoths, and mastodons. As these became stuck in the tar, they attracted predators like sabre-tooths and the American lion, which became ensnared in turn. Birds like vultures, condors, and eagles also tried to scavenge the carcasses and similarly became caught. Plant material blew in. Altogether, more than 660 species are preserved in the tar; there is even one human fossil, a female dating from 9,000 years ago. The mix of species, being heavily biased towards predators and scavengers trying to get at the trapped carcasses, is not indicative of the original community. But preservation of the trapped individuals is outstanding.

Burial and diagenesis

Aerobic (air-living) bacteria play a major role in the decay of organic remains while they are in open soils or oxygenated water. Much has been made in the past of the idea that remains have a better chance of preservation if they pass quickly into a stagnant, anoxic (oxygen-free) environment. Macroscopic scavengers are absent from anoxic soils and waters; in anoxic muds there are no burrowing worms or other creatures to disturb the sediment (bioturbation) or to channel in oxygen. This is the situation, at least seasonally, at the bottom of highly productive lakes and marine lagoons, where there is an over-abundance of organic matter the decay of which uses up the oxygen in the bottom waters. However, it seems likely that the immediate environment of most organisms decaying in wet sediment is largely or wholly anoxic anyway, simply because the aerobic bacteria in the immediate vicinity of the remains use up the local oxygen.

Whatever the conditions under which an organism dies, the remains will eventually be lost unless they become incorporated in the right sedimentary context. Among these are water-borne sands, muds, and clays, and ash falls. The most common of all fossils have simply fallen to (or lived on) the bottom of the sea and become covered by sediment washed in or raining down from above. Freshwater lake deposits like the Middle Devonian Achanarras Fish Bed in Scotland, the Late Devonian Escuminac Bay deposits in Quebec, or the Eocene Green River Formation of Wyoming, where so many superb fossils have been found, often show a series of annual layers (varves), alternating between an organic-rich layer of organic matter that sank to the bottom and decayed due to autumn/winter die-off, and the accumulation of sediment deposited during periods of spring run-off from the surrounding land.

Another favourite case for many palaeontologists is the ox-bow of a meandering river system that becomes cut off and becomes stagnant. Oxygen-deprivation kills all the organisms in it; they sink into the ooze at the bottom until the river floods again and deposits a load of fine mud over them all. Over-bank deposits in a river plain show the same result. In the sea, sometimes an entire living community – a coral reef, for example – may be preserved *in situ* by a tidal wave or submarine slump burying it in sediment. In other catastrophic circumstances, heat and ash from a volcanic eruption may overwhelm whole ecosystems, terrestrial or aquatic.

Most fossils are not found singly but in groups but care has to be exercised in interpreting such assemblages; they may represent a life community or, more likely, they have simply accumulated, after death, in the place where they were preserved. In the case of the Burgess Shale, the organisms seem to have been killed in one environment and then quickly swept into another. The famous Solnhofen limestones of Late Jurassic Bavaria were formed in a series of shallow, hyper-saline, marine lagoons in which there was little life except cyanobacteria and planktonic foraminiferans. The bottom was a fine silt of carbonate-rich mud. As at Rancho la Brea, the fossils preserved at Solnhofen are all forms that lived elsewhere and flew in (*Archaeopteryx* and the pterosaurs), or were washed in. Once they landed in the lagoon, they quickly died and fell to the bottom to be covered in the mud. There was no burrowing fauna and no currents to disturb the remains. The result was a limestone so fine-grained that it was used to make lithographic plates and in which fossils were beautifully preserved.

Burial can cut off the pre-burial phases of decay and dissolution and prevents further disassociation and transport of the remains. The

Environment and preservation

Nor doth it seem consonant to reason, that the part of an Animal Body should so long resist the injuries of so many Years, since we see, that often within the space of a few years the same Bodies are destroy'd totally. But this objection may easily be answer'd by saying, that the whole Business depends from the diversity of the Soil: For, I have seen Beds of a *clayie* kind, which by the thinness and fineness of its juice did resolve all Bodies inclos'd in it; but I have observed *Sandy* Beds, which preserved all Bodies lodged therein.

Steno, *De Solido Intra Solidum Naturaliter Contento Dissertationis Prodromus* (1669; English tr., Hans Oldenbergh, 1671)

remains may, however, still be subject to disturbance by burrowing organisms and bacterial decay will continue. But now the early phases of fossilization are more predictably entrained. Within these burial conditions a new regime of chemical changes begins for both the remains and the surrounding sediments.

Chemistry

Whatever portions of the original organism have persisted this far – perhaps including some organic tissues, more likely consisting simply of some pieces of resistant mineral skeleton – the enfolding of those decaying remains within wet sediment sets up a three-dimensional chemical micro-environment. Here a localized chemical soup is created with special conditions of acidity or alkalinity and a dynamic complex of chemicals some of which dissolve out from the remains into the sediment, and others that enter from the surrounding water and sediment. In most cases the chemical trigger for these changes comes from decay of the residual organic materials in the remains (oxidation of phosphates and nitrates and then reduction of sulphur). A special factor in preservation may be the presence of mats or films of micro-organisms like bacteria over the surface of the remains. These provide an organic matrix for mineralization and trigger the local concentration of phosphates and carbonates. Where this has occurred the level of detail preserved in the fossil may be high.

Because the chemical processes of diagenesis take place in an aquatic solution (the water contained in the pore spaces of the sediment), much depends on the relative solubility of the materials involved. Many substances – tooth enamel, for example – are extremely insoluble in water. Bone mineral is very slightly soluble in water, and more soluble in salt water. Whether minerals actually will dissolve into the surrounding water will also depend on how much mineral is already contained there. For most chemicals, the amount that will dissolve in a given volume of water is limited, beyond a certain point the solution is said to be 'saturated'. Many

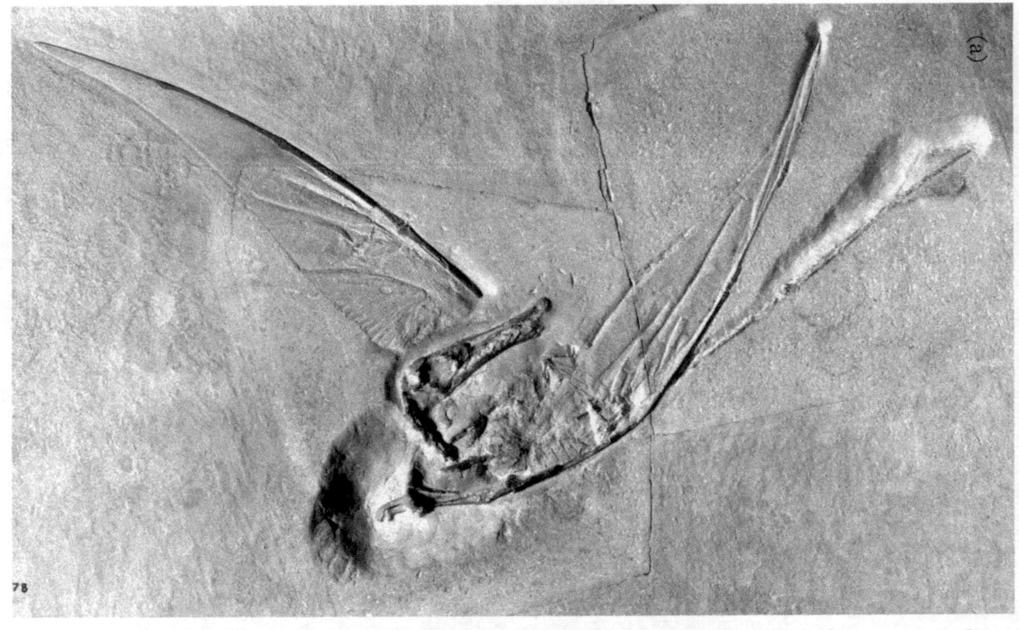

10. Preservation: (a) this specimen of *Rhamphorhynchus*, a flying reptile from Solnhofen, preserves the outline of the delicate wing membrane; (b) under the microscope this cross-section of a Permian fish scale (*Ectosteorhachis nitius*), from Texas, shows preserved blood vessel traces and bone cell spaces, the latter filled with black, iron-rich mineral

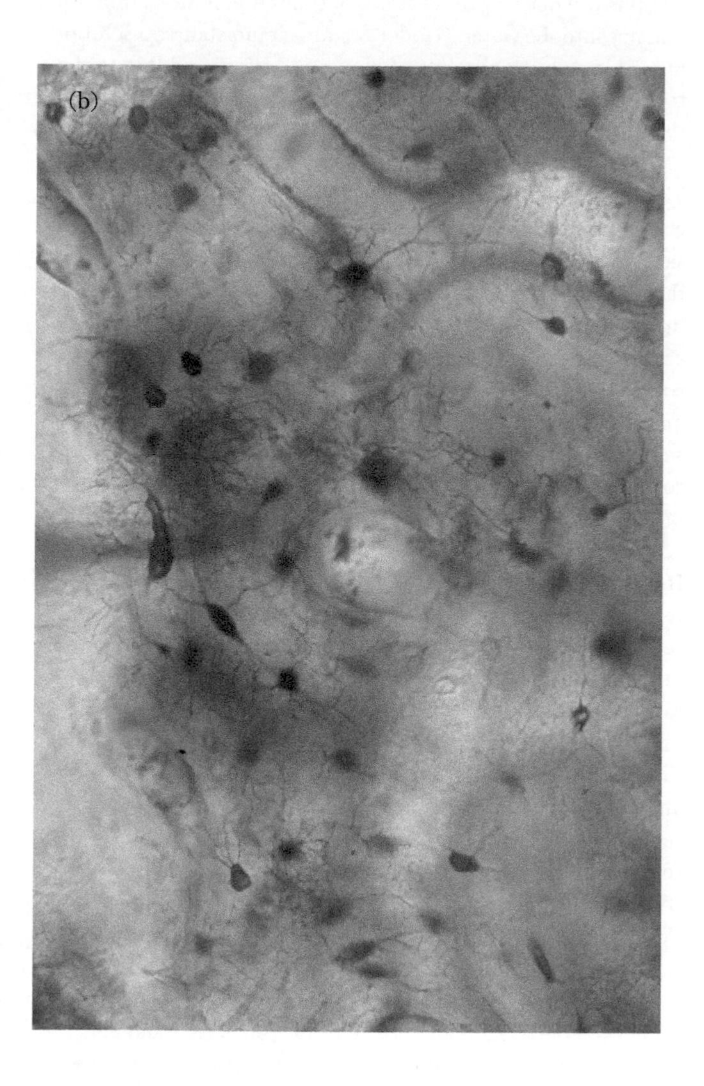
(b)

pore waters are already saturated with phosphates and carbonates and this will delay or prevent dissolution from mineral from the remains into the water. (Under certain circumstances a solution may be in an unstable state of 'supersaturation', and then various triggers will cause the dissolved material to precipitate out again. In that case, mineral is added to the fossil rather than taken from it.)

The skeletons of many invertebrates consist of calcium carbonate; chalk and limestone are made of calcium carbonate. In living skeletons it exists in two different structural forms: *aragonite* in the case of most molluscs, some corals and sponges; *calcite* in most brachiopods, some sponges, foraminiferans, ostracods, echinoderms, and some arthropods; many groups have a mixed composition. Aragonite is rarely preserved intact as it is not stable and is usually quickly replaced by calcite. In the fossilization of something like the shell of an ammonite, once again there is something of a race – between dissolution of the aragonite and precipitation of calcite.

If dissolution proceeds completely before the mineral can be replaced, the fossil may be left as a natural, hollow mould within the sediment – this probably requires the assistance of a coating of micro-organisms. If the aragonite is progressively replaced by calcite *in situ*, the internal microstructure of the shell may be preserved. Occasionally, the original aragonite is preserved and then one may even find the original colours of a Cretaceous ammonite. Both aragonite and calcite may be wholly or partially replaced by dolomite (magnesium carbonate), pyrite (iron sulphide), or by silica in the form of opal or chalcedony.

Typically, vertebrate skeletons are made of calcium phosphates (hydroxy-apatite) rather than calcium carbonate. Bone is a composite material in which the mineral is laid down on a protein framework of collagen and it is typically full of cells and blood vessels. In fossilization, as decay proceeds, the apatite may be replaced with calcite. The microskeleton of collagen typically becomes lost and the spaces fill with re-crystallized calcium carbonate from the pore water. This usually picks up a signature of trace elements like cadmium and chromium, and rare earth elements like thorium and uranium from the pore water. That fact turns out to be very useful because these trace elements can reveal where a fossil came from. For example, fossil fishes from the Late Triassic Newark Supergroup of eastern North America are usually significantly radioactive with uranium. The apatite of bones is occasionally replaced with silica (opal); there is even a whole plesiosaur skeleton from Australia entirely in opal.

The silica skeletons of some sponges, diatoms, and radiolarians are usually in a form (opal-A) that dissolves in alkaline pore water (producing a saturated solution of micro-crystalline opal-CT) and is later re-crystallized as quartz. Before that can happen, however, the original silica skeleton may dissolve away completely or be replaced by calcite or, in sediments rich in organic material, by calcium phosphate.

In the process of permineralization, dissolved minerals in the pore water of the sediment permeate through, and precipitate within, all the porous spaces left in tissues such as wood and vertebrate bones. The shells of invertebrates lack such spaces. Plant material is more commonly silicified than animal tissue, even to the extent of whole petrified trees. In special cases like the Early Devonian Rhynie Chert of Scotland, an environment of shallow freshwater pools was periodically engulfed by water from hot springs saturated with silica. Here, organic decay was cut short and very rapid silicification allowed the preservation of soft tissues. Other kinds of permineralization involve the formation of pyrite and various phosphates. The sooner permineralization occurs, the better the chances of preservation of details of soft tissues.

Perhaps the most obvious example of the chemical action within the mini-environment surrounding buried remains is the formation of nodules. Here, a nodule of carbonates forms within the surrounding

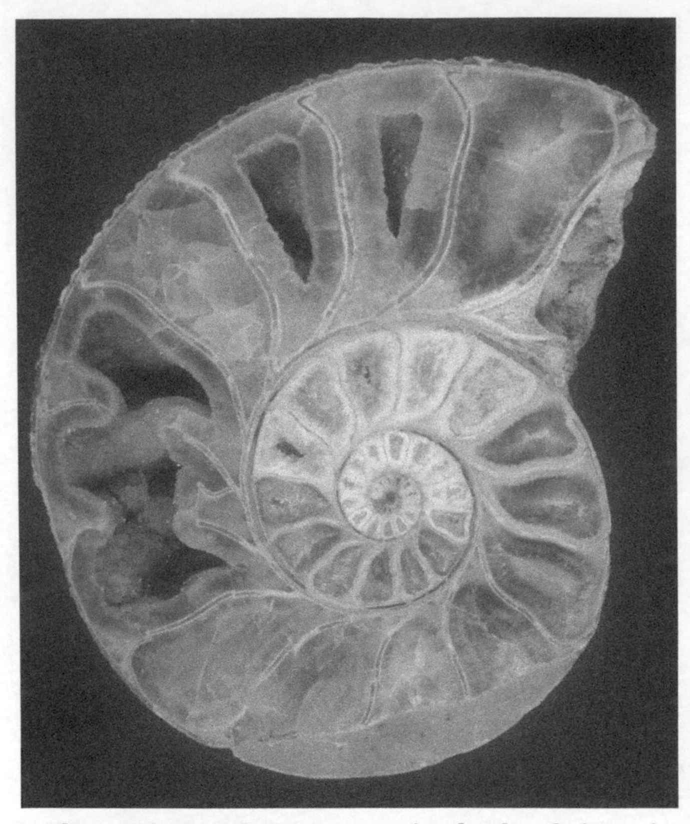

11. The Jurassic ammonite *Lytoceras*, sectioned to show the internal structure; the original air chambers have been filled with calcite, some of which has subsequently dissolved away

sediment around a nucleus of organic material, rather like a pearl growing around a grain of sand inside an oyster. The shape of the nodule and the details of its formation depend acutely on the position of the remains relative to the zone of new sedimentation and factors like methane production in the breakdown of the organic materials. Concretions are found in many sites around the world. In most cases, the nucleus was only a fragment of organic matter, but occasionally a concretion will turn out to contain a superb fossil. Among the more famous places for nodules are the Devonian Gogo Formation in Western Australia, the Pennsylvanian (Late Carboniferous) Mazon Creek locality in Illinois, the Triassic nodule beds of Madagascar, the Cretaceous Santana Formation of Brazil, and Fox Hills Sandstone of North Dakota, and in the Eocene London Clay. In nodule formation, it is thought that reduction of the organic material of the nucleus results in local supersaturation of the pore water with carbonate. This carbonate is then re-precipitated around the periphery of the developing nodule (as the minerals siderite or ankerite).

Despite the general decay of the organic materials in developing fossils, in special cases it is interrupted so that, as in the formation of peat, coal, and oil, much of the original organic material is preserved. Coal is created by the incomplete decay of vast peat-like deposits of plant material that accumulated in ancient tropical and subtropical swamps. Oil and gas represent the remains of hydrocarbons sequestered in the bodies of countless trillions of planktonic micro-organisms from ancient oceans. Gas and oil migrate through and accumulate in porous rocks, from which they can be harvested. In all three cases, heat and pressure are essential to the process.

Despite all these vicissitudes, some proteins – collagens, for example – may endure for a long time in fossil remains. Amino acid signatures may persist for several millions of years (up to 100 myr has been claimed), and even fragments of DNA sequences may survive for 100,000 to 120,000 years. The larger the molecules that are preserved, the more information they potentially contain. On the other hand, in some fossils all that remains is a carbon film, like a photograph faithfully recording the shape of the organism, on the rock.

Becoming rock

Diagenesis of the fossil remains is accompanied by a whole series of parallel processes of lithification (rock-making). Compression and

Fossils and industry

Fossils have a number of directly economic uses, beyond their contribution to pure science. Rocks made of microscopic fossil diatoms are mined for a wide range of uses, such as in filters (diatomaceous earth) and even in cosmetics. Coal, gas, and oil are 'fossil fuels' and the product of once-living organisms, although there is a growing argument that some methane gas might have been formed from the rocks themselves under great pressure. They are all organic hydrocarbons, of which the principal source is dead plant material – in the case of coal, from ancient freshwater peat deposits laid down in swamps; in the case of oil, from microscopic plankton living in the sea.

Just as 'signature' fossils allow geologists accurately to relate the layers in a stratigraphic sequence, those same signatures can be used predictively. The floors of all ocean basins, from ancient times to the present day, accumulate layer upon layer of microfossils: foraminiferans, coccoliths, diatoms, and radiolarians. Chalk deposits, thousands of feet thick, from Europe to Australia, testify both to the productivity of the oceans and to the resistance of these microskeletons to decay. Today, prospectors for the gas and oil industries use the characteristic patterns of these microfossils when analysing the cores from boreholes. In that way, it is possible to read the geology even from cores drilled from a ship, and to predict not only where reservoirs of oil will be found, but how much oil they will contain.

removal of the water encourage the cementing together of sediment grains and crystals of new minerals such as feldspar may be formed. The extent to which the fossil remains are deformed by compaction of the rocks depends very much on the relative timing of events. If lithification sufficiently precedes compaction, the fossil remains are preserved three-dimensionally intact. Very many fossils end up wholly or partially flattened, however, and are often also curiously distorted through plastic deformation of the fossil and rock together. Heat and pressure play a major role in late diagenesis, but not in every case. Some really old fossils look quite natural, others have evidently been cooked.

Opinions vary as to the total length of time needed to 'make a fossil'. The processes of chemical stabilization and diagenesis must begin within days of burial and permineralization may be largely complete within tens of years. For the sort of preservation of fine details, and even soft tissues, that make the so-called Lagerstatten deposits so important, the early stages must have happened very quickly. Even there, however, full consolidation and lithification of the enclosing sediments may take thousands or millions of years.

Trace fossils

While fossil remains of hard body parts and occasional soft parts tell us a great deal about the original organism, trace fossils (ichnofossils) give us a different, parallel, view into the extinct world. The phenomenon includes track-ways, burrows, faecal remains, borings, and tooth marks that, together with related phenomena such as raindrop impressions, mud cracks, ripple and current marks, provide unique environmental and behavioural information.

Most trace fossils were created within a soft substrate or when an organism made a mark on a surface with the right properties, wet mud being the most obvious. The surface must not have been too wet, or the trace would quickly be lost, or too sticky, in which case the definition of mark would be obscured. Perhaps because they are usually very large and deep, dinosaur track-ways are quite common. Very rarely, of course, can the organism that made the trace be identified as to species; that is only possible under the ideal situation that the animal is found lying dead in its burrow or at the end of the track-way, as in the famous case of a horseshoe crab washed into the Solnhofen lagoon that literally dropped dead in its tracks.

Examination of a vertical section through a track-way usually shows that the imprint may be pressed through several thin layers of sediment and the bedding plane that splits to reveal the track may not be the 'top' plane marking the separation between the original surface and some later deposited sediment. This means that, for a heavy imprint, the original impression of a track-way may be preserved in the deeper sediments even if the actual mark on the exposed surface had been disturbed or washed away.

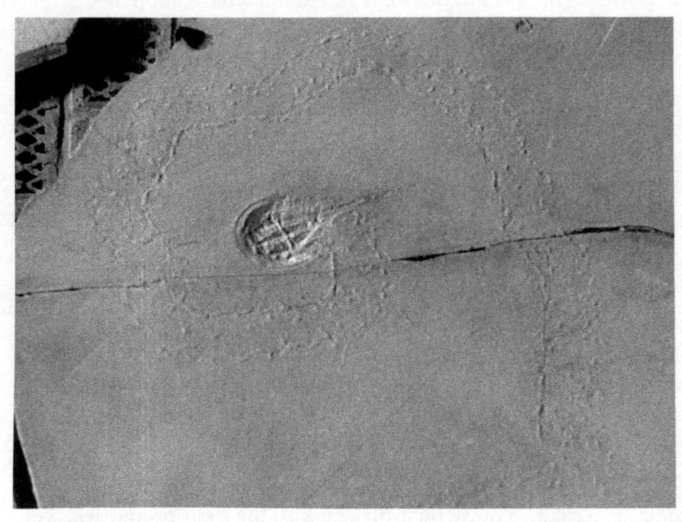

12. This horseshoe crab, closely related to the modern *Limulus*, crawled onto Solnhofen's Jurassic toxic limestone mud and quickly died, literally in its tracks

Discovery and preparation

The remains of an organism can be lost at any stage and, even when preserved, fossils may be surprisingly elusive – buried by the accumulation of so many layers of younger rocks that they never appear on the surface at all. The rocks in which they were entombed may be destroyed through erosion or subduction. Only when a particular fossil bed has once again become exposed at the surface (the overlying rock layers having been removed by nature or by man) can it be found and collected. In Western Europe where the mild climate and (mostly) low altitudes ensure abundant growth of vegetation, discovery of fossils depends on chance exposure along stream beds, steep hillsides, sea cliffs and in quarries, road and railway cuttings, and mines. By contrast, in places like western North America, Australia, and the Gobi Desert a relatively bare landscape erodes more quickly and reveals more fossils – if they are there. Wherever fossils outcrop, they have to be found relatively quickly before they are reduced once again to grains of sand by weathering and erosion.

Finding the right places and the right rocks to explore is both a science and an art. Often, discovery of a fossil site depends first on the basic geological exploration and mapping having been done by someone else. Many important sites have been discovered first by someone finding a scrap or two of fossil weathering out onto the surface. Ultimately, finding fossils depends on two things – wearing out a lot of shoe leather and having 'the eye'. While the former is obvious, the latter is not. The great fossil collectors have always had the capacity to see fossils where someone else would walk right past. And it appears not to be something that can easily be learned. Not having the eye myself, I discovered early on that my best collecting was in the drawers of museums where a hundred years of work by experts was concentrated. But those who want to study something like the composition of a fossil community have to go out and do the hard work themselves, because it rarely turns out that others have collected with the same purpose as oneself.

Even when a fossil has been found, carefully collected, and brought home for study, much remains to be done. In the laboratory, the fossil is taken over by a 'preparator' on whose technical skills much depends. The protective wrapping must be removed very carefully because the rock will have dried out in the time since the specimen was collected. Any glue or hardener applied in the field must be removed. Then the specimen is carefully prepared, usually in a time-tested mechanical way in which the enclosing rock is removed almost grain by grain under a microscope with tiny steel chisels and needles. Sometimes mechanical tools like dental drills can be used if the fossil will stand the vibration. Occasionally, the surrounding matrix is removed by etching it with acids. If the fossil is made of something relatively resistant like silica and preserved in limestone, the limestone can be carefully dissolved in dilute acid, leaving the silica fossil behind. This sounds like a relatively simple process, but it is always difficult and time-consuming as the fossil itself has to be protected with waxes and lacquers while the whole thing is submerged in an acid bath. Then it is washed and the whole process repeated many times.

Some fossils end up in sediments like chalk and lignite that are so soft they can be dug out with a palette knife and cleaned with water and a brush. Other rocks are so tough, especially when cemented by iron minerals, as to be impervious to the toughest chisels and all but the strongest acids. On average, the younger the sediment, the better the fossil and the easier it is to 'prepare'. The least invasive way to examine a fossil is by CAT scan, reconstructing the whole animal by computer. The most invasive is to grind serial sections at very fine intervals (for example, 30 microns), photographing each time, and then to make a reconstruction using the computer. In any case, almost invariably, many days and even weeks will go by before the full details of the specimen are revealed.

Chapter 6
Bringing fossils to life

Palaeontology is rarely a glamorous saga of digging up whole skeletons or entire communities preserved intact in a nice soft rock; most often it is hard graft – picking over thousands and thousands of fossil fragments in order to find tiny clues that can be assembled, often over years, into a coherent pattern. A great deal of the fascination of palaeontology therefore exists in the role of the scientist as detective. A palaeontologist works in a similar fashion to the familiar forensic scientist of detective fiction (and real life), teasing out the mysteries of a single life and a whole world from a few inanimate remains.

Palaeontology has several extraordinarily ambitious goals. Historically the first of these was to use fossils in the science of stratigraphy, and to discover and interpret the structure and history of the earth itself. Each layer (stratum) of the sedimentary parts of the earth's crust has its own characteristic fossils, giving a clue to its age relative to other strata, and also the original environment. As the strata, through time, show different *but obviously related* faunas and floras, next came the second goal: to re-assemble an all-encompassing genealogy – the evolutionary inter-relationships – of fossil and living organisms, back to the origins of life itself. No palaeontologist would ever imagine that the resulting family tree could ever be complete. But every fossil that is found has the potential to add something to the evolutionary database. The third major task of palaeontology is more ambitious still: to understand what the fossil organisms were like in life – not just to identify them, but to reconstruct their appearance, their mechanical and physiological functions, their ecology, and even their behaviour.

The living organism

Reconstructing the living organism depends in large part upon having a thorough knowledge of the biology of living organisms and finding an appropriate analogue among living organisms. We can, for example, make inferences from the teeth of a fossil about what the animal ate; the jaws can tell us how it chewed; joints, muscle scars, and limb proportions tell us how it walked or ran; the shell can tell us how it burrowed; from the proportions we can estimate the size of the whole organism. With many animals, it is often possible to tell whether the individual was a male or female. In vertebrates, fused sutures in the skull indicate an adult, and so on. The shapes of leaves of living trees are characteristic of the environments in which they live: dissected leaves with long points tend to be from plants that live in moist environments, long thin leaves are most efficient at gathering the sun's energy in dry, open environments, rounded leaf margins are more characteristic of temperate conditions. At the same time, our reconstructions of new fossils (and indeed our constant revision of reconstructions of old ones) depend on, and are skewed by, the living models we choose to compare them with. Obviously, the fossil animal or plant we are trying to reconstruct probably looked – and lived – more like its closest living relatives than to anything else.

Armed with these principles, even if we have a fossil of one part of the skeleton we can make enlightened estimates of what the contiguous parts must have been like. Something with the front legs of an elephant will not have had the hindlegs of a deer, for example. In the early 19th century, Professor (eventually Baron) Cuvier in Paris was so sure of his ability to predict the whole skeleton from just a small part that he developed a 'principle of correlation of forms in organized beings' that codified this kind of consistency. He was sure that he could reconstruct a whole skeleton from just a single bone.

A great deal, therefore, depends on the palaeontologist knowing what sort of organism he or she is dealing with. A fossil mammoth, for example, can be expected to have a lot in common with a living elephant species. When we deal with a group of giant fossil organisms with no obvious close living relatives, however, the problems become more difficult. The dinosaur *Iguanodon* was so named because its teeth were similar to those of a modern lizard, the iguana. But a dinosaur is not just an overgrown lizard. Nor are ichthyosaurs – as they were first thought to be – crocodiles. Plesiosaurs (which have been described as looking like a snake threaded through a turtle) were even harder to interpret. When the first plesiosaur was discovered in 1821 by Mary Anning, Cuvier firmly concluded that it must be an error or a fake, because no animal was known with such a small head and long neck and body. Such an animal could not exist! It turns out that in palaeontology, as elsewhere, and for Cuvier just as for all of us, the most difficult thing to know is the extent of your ignorance. The fossil record has thrown up many organisms that, on the face of it, ought not to be able to exist. There are many fossil organisms for which there is no living analogue. And all analogues are simply that – the basis for a reconstruction rather than a proof of identity.

As discussed in Chapter 3, the discovery of dinosaurs in the 1820s to 1850s launched a major phenomenon in popular culture. When William Buckland at Oxford University first described *Megalosaurus* from some pieces of jaw, a pelvis, and a few limb bones, no comparable bipedal animals were then known and it made sense to reconstruct it as a quadruped – part frog, part hyena. Joseph Leidy in Philadelphia reconstructed his *Hadrosaurus* of 1858 as being bipedal, but it was only when a group of quite complete skeletons of *Iguanodon* were discovered in a coal mine in Belgium that palaeontologists had irrefutable evidence that they stood on their hindlegs and to this day, one cannot be really sure what the tiny forelimbs of something like *Tyrannosaurus rex* were used for (although there is plenty of speculation).

At one time, the huge herbivorous sauropod dinosaurs like *Apatosaurus* were assumed to have been too heavy to live on land, but must have lived in lakes with their weight partially supported by water. They were interpreted as slow and stupid animals, as were all dinosaurs: cold-blooded in physiology and small of brain. Scientists boldly announced that the animals were so long that nerve impulses would take too long to pass from brain to tail, and an accessory brain must have existed in a hollowed-out region of the vertebral canal in the sacral region. The bipedal dinosaurs were seen as almost equally clumsy, with their long tails dragging in the ground or acting as a third leg when the animal was at rest.

In the 1960s and 1970s, the study of 'form and function' in living animals – a combination of biomechanics, physiology, behaviour, and ecology – entered a new experimental and analytical phase. Every aspect of biology of living organisms, simple or complex, plants, invertebrates, or vertebrates, was opened to new scrutiny and the results quickly applied to fossils. With dinosaurs, new studies of the sizes of muscle insertions on the skeleton, and the precise angle of the hip and shoulder joints, coupled with detailed experimental studies of the physiology and locomotion of living reptiles (notably aided by motion picture X-rays), started to tell a new and different story.

A crucial element in this was the description, in 1969, of a medium-sized Early Cretaceous dinosaur from Montana, by John Ostrom of Yale University. In many respects it seemed a typical small carnivorous dinosaur, except that it had a huge, vicious-looking claw on each hind foot. The tail was supported by a long series

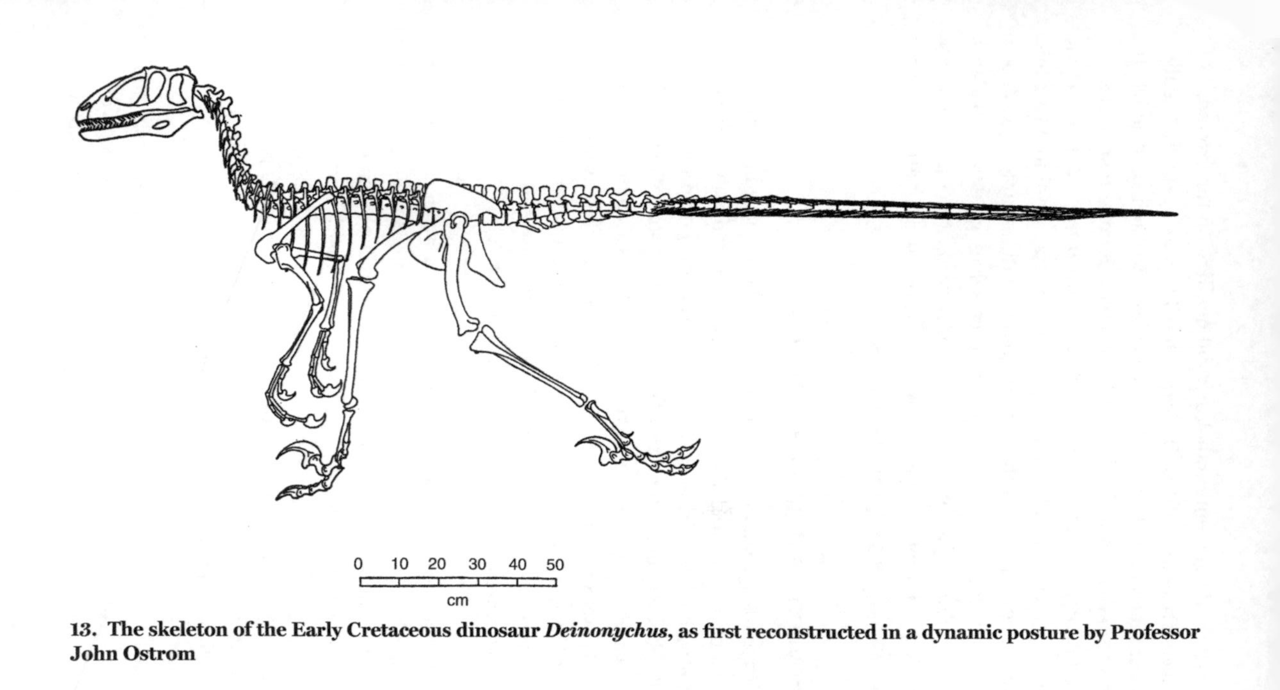

13. The skeleton of the Early Cretaceous dinosaur *Deinonychus*, as first reconstructed in a dynamic posture by Professor John Ostrom

of ossified ligaments and clearly did not drag on the ground. Reconstructions of *Deinonychus* ('terrible claw', as he eventually called it) revealed a rather fearsome predator, agile and active. It must have been warm-blooded and smart. Similar discoveries followed, including forms like the tiny *Velociraptor*. And with that a whole new style of thinking about and portraying dinosaurs came into vogue. The concept of warm-blooded dinosaurs, and a flood of new information about the fossils themselves, gave birth to a new generation of full-blooded, hot-blooded reconstructions, sometimes erring as far intemperately towards the overly dramatic as older views had been clumsy and lifeless.

Environment and behaviour

Often fossil remains inform us directly, both about how the organisms died and also of something of how they lived. In addition to tooth marks giving evidence of scavenging and predation, skeletal remains often show fracture healing and evidence of diseases like arthritis; Pleistocene cave bears seem to have been quite creaky in that respect. The *Tyrannosaurus* dinosaur specimen at Chicago's Field Museum, nicknamed Sue, had successfully survived several broken ribs, lesions of the jaw, and osteomyelitis (inflammation of the bone tissue) of the left leg.

There are fossils of ichthyosaurs with the stomach contents intact, and others that died in the act of giving birth. Numerous fishes have died trying to eat another fish that was too large. Insects in amber are sometimes preserved in copulation. Parasitic and boring organisms leave a number of physical traces in their prey. Nests of the dinosaur *Maiasaurus* contain young, suggesting parental care. There is a whole sub-science concerning 'coprolites' – fossil excreta – a field full of dietary information.

Trace fossils are particularly useful. At the Middle Jurassic (168 myr old) Ardley quarry near Oxford, an ancient shoreline records dinosaur tracks stretching over some 200 metres. A group of several huge sauropods had ambled slowly across a limey-mud flat, steadily putting their hind feet onto the dinner-plate-sized spots where the forefeet had just trodden. Another set of tracks, heading in the same northwest direction, was made by three or four bipedal theropods – 20- to 30-foot-long carnivores (probably *Megalosaurus*). Using a simple formula relating the height of the hip joint from the ground and the length of the stride, it is possible to calculate that the sauropods were walking slowly at less than 2 miles per hour. Most of the time the carnivores were travelling at a 2.5 to 3 miles per hour walk. Then they broke into a run (at about 8 miles per hour), for a short while – but well below what would have been their maximum of about 18 miles per hour. It is tempting here to reconstruct this as the small herd of giant herbivores being trailed by those megalosaur predators and to imagine that just 'off-screen' (further along the exposure where things are hidden by an under cliff) there was terrible carnage. But equally, the megalosaurs might have gone along the beach first, and even been chased off by the approaching giants. There is also the very dull possibility that the two groups wandered along the same beach at different times and never saw each other. One has to be careful not to be over-imaginative about trace fossils!

A lot of the behaviour that is depicted in paintings and reconstructions of fossils is based on rigorous functional analyses of structure. Some, particularly in the animations created for television and films, is conjectural and/or based on assumptions about living models. For example, the animated *Apatosaurus* in the BBC's acclaimed series *Walking with Dinosaurs* started with a walking and running elephant. It provided the dynamic (in this case, electronic) equivalent of the sculptor's armature, onto which the long neck and tail of the dinosaur were been grafted graphically. The effect is superb, but once you know there is an elephant in there, it is harder to see the *Apatosaurus*.

Most fossils are not found alone but in multiples. These collections should be termed 'assemblages' to distinguish them from the

14. Plot of track-ways in Ardley quarry, Oxfordshire. A group of giant herbivorous sauropods (solid lines) crossed a muddy Jurassic beach; whether they were aware of the two carnivorous megalosaurs (dashed lines) is a matter of speculation. Both groups were heading slightly east of north

15. While the juxtaposition is dramatic, the maker of these human footprints did not meet the jaguar that also crossed this Costa Rican river beach

original communities of which they are a reflection. The natures of these associations are highly informative. Many kinds of plant fossils are precise indicators of climate or even, for modern flowering plants, the altitude at which they grew. The sequential layers of sediment and entrapped fossils in lake deposits allow us to read the patterns of climatic change over hundreds or thousands of years. Despite many difficulties, fossils can be used to reconstruct a great deal of palaeo-ecological information concerning the ecological niche of the individuals, the community made up of all the individuals, and also the broader habitat.

Sometimes the simple geographical distribution of fossils presents a challenging ecological problem. Large dinosaurs like the carnivore *Troodon* and the herbivore *Edmontosaurus* are found in the Cretaceous of Alaska; but most people think that those regions were polar or sub-polar at that time. Even if the temperatures were more moderate up north in those days, how did dinosaurs survive the long, dark winter months? It is thought that they did not migrate like modern caribou. Is there some explanation that we just haven't spotted yet?

Fossils and artists

Palaeontologists have always depended a great deal on the skills of artists to portray their work. One of the factors in the acceptance of Robert Hooke's ideas about the true nature of fossils in his *Micrographia* (1665) was his superb drawings of the specimens (as a youngster, Hooke trained with the London artist Peter Lely). Not all palaeontologists are gifted enough to make the final reconstructions of the fossils they work on, and very few fossils are complete as single specimens. An artist is usually necessary to make a single whole out of the parts. At its best, the collaboration between scientist and artist is truly synergistic.

The role of the artist is twofold. First, the artist may be asked to create, as faithfully as possible, an illustration, almost photographic in detail of the specimen. Second, the artist may be asked to portray a theory about the fossil. That a particular fossil animal had long legs, or blunt teeth, or that a plant had pointed leaves, is a matter of fact; how they functioned in life is another. In many instances, the distinction between a faithful depiction of what exists and a reconstruction of what might have existed becomes blurred, although it is incumbent on all concerned to make sure that any assumptions and hypotheses in the final representation are fully explained. One rule for the reader generally applies: illustrations for scholarly monographs are more factual, those for the popular market are often more conjectural.

Perhaps the most dramatic use of the artist's skill is in the reconstruction of entire ancient landscapes complete with their animal and plant denizens in full glory. In museums these landscapes were in the past often translated into full-sized or miniature dioramas – three-dimensional works in which the depiction of the fossil organisms is the work of the sculptor rather than the painter, while the latter is more than fully employed in depicting the background setting. Perhaps the first attempt to paint a whole landscape was by the geologist Henry de la Beche, a skilled draftsman with a flair for the dramatic. His 1830 watercolour and ink reconstruction of the Dorset ichthyosaurs, plesiosaurs, and pterosaurs collected by Mary Anning and others, set within a Jurassic marine landscape, had within a decade become the model for dozens of imitators and spawned a genre of scientific-commercial art that continues to grow and develop.

The animals in de la Beche's ancient Dorset may seem wooden, but the effect is still one of high drama and violence. It is not a world into which humans would want to venture except vicariously. (None of the many excellent reconstructions of other subjects, such as invertebrate life in the Palaeozoic, produce such an effect.) The tradition has continued ever since, often creating a dynamic tension between the desire to show something dramatic and the need to remain faithful to the sober underlying reality. In this respect,

16. Henry de la Beche made prints of his 1830 painting *Duria Antiqiour* ('Ancient Dorset') for sale at two and a half guineas, with the proceeds going to support Mary Anning

17. Rudi Zallinger's giant fresco in the Peabody Museum at Yale University set the standard for modern reconstructions of ancient life. This section shows a Permian landscape, including mammal-like *Dimetrodon* with its extraordinary dorsal 'sail'

among the many talented portrayers of dinosaurs, perhaps no modern artist has produced such an array of brilliant and dramatic work as the cartoonist Bill Watterston in his *Calvin and Hobbes* series.

Benjamin Waterhouse Hawkins (Chapter 3), in addition to his dynamic sculptures of dinosaurs, painted some large murals of Mesozoic life at Princeton University around 1870. Early in the 20th century, the art of depicting whole habitats took a step forward in the work of Charles Knight, who painted huge murals of great life and accuracy in the museums at New York, at Chicago and Los Angeles. In perhaps the last of this genre, in the 1950s Rudi Zallinger painted a 110-foot-long fresco of the Age of Reptiles (followed later by the Age of Mammals) at the Peabody Museum, Yale University. This work, distributed as posters to almost every school in America and many worldwide, summarized the very latest in what was known about the appearance and behaviour of Mesozoic animals and plants. It is a superb work, careful and accurate, although the poses of some of the dinosaurs now seem conservative and static in comparison with modern ideas. As a portrayal of ideas as well as facts about fossils, it stands as a benchmark to be compared in the long line of other such efforts, starting with de la Beche, the same way that we compare the fossils themselves.

Some fossils show evidence of patterning in the integument, but the colours used in any depiction of a fossil, like much of the posture and behaviour, are partly from the artist's imagination and partly guesses from the biology of living creatures. Although there is an argument that, if dinosaurs are related to birds, they might have used colour in their behaviour, we cannot know about warning or protective colorations, hair patterns, or iridescence in long-extinct creatures – let alone mating behaviour or ritual displays. All big dinosaurs must surely have had thick skins, and thus were most likely a disappointing dull grey colour like an elephant or rhinoceros.

Chapter 7
Evolving

Evolution can be a contentious subject when people do not understand that the term can be used in several different contexts with significantly different meanings. Fossils have played a major role in the development of all aspects of organic evolution and continue to play an essential role with respect to working out the patterns and causes of evolutionary change. First and foremost, fossils are the documentary evidence of evolution in the sense of change in life over time: early simple (lower) organisms being supplemented by and often supplanted by more complex (higher) forms. The fossil record gives us a series of consistent patterns of these changes with time, involving inexorable processes of origination and extinction, constant faunal and floral turnover, and it is a reflection also of the fluctuations in the environments in which they all lived. The patterns themselves are contingent; at any one time in geological history, what exists is a function of what existed before. Their consistency forms the basis of the use of fossils in stratigraphy.

Our understanding of patterns of changing diversity in the Phanerozoic record has stood the test of time well. Even though the details may be revised every time a fossil collector sets to work, the general features are always confirmed. Among the vertebrates, the fishes came first, then amphibians and reptiles; birds and mammals came last, humans last of all (so far). There are no Ordovician mammals, Devonian dinosaurs, or Jurassic humans. None of the huge reptile groups of the Mesozoic, such as ichthyosaurs, survived into the Tertiary. There are no living plesiosaurs (the supposed Loch Ness monster notwithstanding). We can argue about what processes might have caused all these changes, but the raw data (the fossils and their relative ages) remain, not as hypothesis but as fact. If we were ever to find a human fossil incontrovertibly nestled in the arms of a dinosaur (something that comic strips regularly envisage), our whole concept of the evolutionary process over time would be almost irremediably negated. (There have been flawed attempts by anti-evolutionists to identify human track-ways alongside dinosaur tracks at the Early Cretaceous site in Glen Rose, Texas.) The fossil record is perhaps one of the most constantly tested 'facts' in science. Every day, somewhere on this earth, a palaeontologist is digging up a fossil. Often these fossils allow us to refine our view of some part of the great evolutionary pattern of life, but they never overturn it.

The second major meaning of the term 'evolution' involves relationship – the concept that all the different kinds of organisms, living and fossil, can be placed (on the basis of their structure and DNA where possible) in a single complex scheme of similarity and difference. Blackbirds and thrushes are more similar to each other than either is to woodpeckers, and so on. Evolution explains these patterns as representing genealogy. Instead of forming a series of independent lines, all living and fossil organisms are related to each other in a pattern of diversification by branching that is only consistent with relationship. Evolution explains that this diversification has resulted from a process of descent from common ancestors. Fossils then represent a series of snapshots (but not yet the complete movie) of a vast family tree, a portrait gallery of all life.

By the time Charles Darwin came on the scene to provide the third meaning of evolution in the shape of an actual mechanism – natural selection – evolution in the previous senses was already quite familiar. And when someone says that 'evolution is only a theory', they are referring to natural selection. Evolution as organic change over time is a fact.

More gap than record?

As a science, palaeontology must be grounded in an excellent understanding of the structure and identity of fossils. When we say that something is rare, or another is common or widely distributed, if we conclude that some phenomena evolved at a faster rate than others, we need to be sure that our sampling is complete and quantitatively rigorous.

The concept of continuity is at the centre of all palaeontological research. In principle, the geological record should present a complete, graded, year-by-year fossilized account of life on earth. In practice it does not. In lake deposits one may find true year-by-year deposition of fossils extending continuously over a few hundred or as much as thousands of years, but not millions. For example, the fossil lake deposits of the Newark Supergroup of eastern North America provide an excellent record of continuous change in fish faunas at the end of the Triassic.

Charles Darwin, for one, realized that the stratigraphic record and its fossils, while they had the potential (now largely realized) to document the overall course of evolution, would perhaps never provide the smoking gun of evidence concerning how transmutation of single species actually occurs. We have only an incomplete record of the history of the earth and life on it, detailed here, patchy there, missing elsewhere. Huge chunks of the rock record no longer exist, having been destroyed in the inexorable geological processes like glaciation and plate tectonics. Other sequences have not been found, or are buried so deep they will never be exposed. The late Derek Ager, a somewhat iconoclastic British geologist, summed up the situation in a memorable phrase, describing stratigraphy as: 'more gap than record' and, like a child's description of a net, a lot of holes held together by string. It means that we must use the record with caution, paraphrasing Hamlet – 'there are more things in heaven and earth than are dream't of in our palaeontologies'.

The total number of described fossil species, over the entire Phanerozoic, is about 250,000. As there are between 1.5 and 4.5 million species living today and the Phanerozoic has lasted 545 myr, the problem is obvious: by any measure, the fossils that we know represent only a small proportion of the total biological diversity that has ever existed. As discussed in Chapter 5, certain kinds of environment will be much less likely to yield a significant fossil record than others. We have a poor record of upland, high-energy, depositional environments, and most tropical environments, for example. Certain kinds of organisms are unlikely to be fossilized. (The fossil record is particularly poor with respect to the insects, which today make up more than 50% of all the macroscopic-sized animals on earth.)

While we cannot possibly know more than a tiny proportion of the

Imperfection of the record

Just in proportion as ... extinction has acted on an enormous scale, so must the number of intermediate varieties, which have formerly existed on the earth, be truly enormous. Why then is not every geological formation and every stratum full of such intermediate links? Geology assuredly does not reveal any such finely graded organic chain; and this, perhaps, is the most obvious and gravest objection which can be urged against my theory. The explanation lies, as I believe, in the extreme imperfection of the geological record.

Charles Darwin, *On the Origin of Species* (1859)

species that have ever lived, if we confine ourselves to groups likely to be preserved in the record, such as hard-shelled marine invertebrates, the situation is quite good. At the higher categories such as families and orders, the record may be 70–90% complete; at the level of genus it might be 50–60% or less. A number of sophisticated quantitative approaches can be used to estimate just how incomplete (and therefore how useful) a particular portion of the fossil record is. We can, for instance, measure the rate at which new species or higher groups are being discovered in relation to research effort. For example, at the early stages of collecting at a major locality like Solnhofen, the number of new species described may grow in direct relationship to the number of specimens collected. But soon the proportion of new species discovered will tail off. Discovery of new genera declines even faster, and of new families and orders, faster still.

This question of completeness is important because of the potential that the fossil record has to reveal not just the qualitative nature of evolution, but also quantitative features such as rates of evolution and the patterns of diversification over time. In fact, however, the critical question is not the completeness, per se, of the record, but its adequacy to answer particular questions. The realistic view is that, while we will never fully quantify the entire extent of global biodiversity in the fossil record, we can make quite accurate studies of portions of the record and of particular kinds of organisms and certain categories of environment.

All of which means that, as in every kind of science, defining the question in the first place is most important task of all. Then the data sets can be matched to the question. For example, study of one particular set of fossil beds may answer a question about evolution of sessile marine organisms over a given time period – perhaps revealing large scale extinction. But that will probably say nothing about what might have been happening in a contemporary freshwater lake, where speciations might have been at a high. If we simply averaged what was going on in the two settings, we would

Human fallibility

The study of organic remains is beset with two evils, which, though of an opposite character, do not neutralize each other as much as at first sight might be anticipated: the one consisting of a strong desire to find similar organic remains in supposedly equivalent deposits, even at great distances; the other being an equally strong inclination to discover new species.

Henry de la Beche, *A Geological Manual* (1831)

come to the erroneous conclusion that evolution was proceeding at some 'normal' rate. On the other hand, if the pattern of change was the same in two different ecological settings simultaneously, then we would have discovered the operation of some more global phenomenon.

Global Phanerozoic diversity

Ironically, the most widely discussed quantitative studies of fossil diversity concern a question that is most difficult to resolve precisely: how have the total numbers of species on earth changed over the past 545-myr span of the Phanerozoic? At the beginning of life on earth there were obviously fewer different kinds of organisms than there are today. It is therefore natural to ask: is total diversity still increasing and, if so, why? Or has some kind of upper limit (analogous to what ecologists call 'carrying capacity') long since been reached? Are there recognizable patterns of fluctuation in rates of origination of new species and particularly, in the rate of extinctions? Where global diversity has increased, has that occurred through increases in the rate of diversification, or have extinction rates decreased? Or both? How, if at all, are originations and extinctions causally linked?

An even more difficult question to answer would be: has total biomass changed in any particular pattern or patterns over time? That is, when in the past there were fewer *kinds* of organism, were there were more *individuals* occupying the same total of all ecological space?

These are fascinating and important questions, especially as we contemplate the possibility that the last 10,000 years have seen a drop in global biodiversity. The answers are of interest to astronomers thinking about the history of the solar system and beyond and to environmentalists worried about the rate of modern, human-driven loss of biological diversity.

Analysis of global diversity over time has the attractive quality that it can be summarized in a simple graphic. When one plots the numbers of different kinds of fossil organisms that have been discovered in the stratigraphic column (using genera as a surrogate for species, and concentrating on hard-shelled, sessile, marine invertebrates), a fascinating picture emerges. Ever since this was first done by John Phillips at Oxford in 1860, global diversity has appeared to have increased (unevenly) over time, with more genera and families of organisms (and by extension more species) living today than ever in the past.

If this picture were true, it would be extremely exciting – revealing the living world very much as a work in progress, rather than holding at some steady state attained in the distant past. And that would launch us into testing a range of possible causal factors. It might mean that the nature of the various taxonomic categories (species, genus, family) is continuing to change over time – each becoming more and more restricted, allowing the existence of more different kinds. It might mean that the habitable parts of the earth have increased in extent. Or that given environments have become more and more sub-divided, with more and more different kinds of habitats, each supporting more and more diverse (specialized) species. Or that each different major kind of organism that has

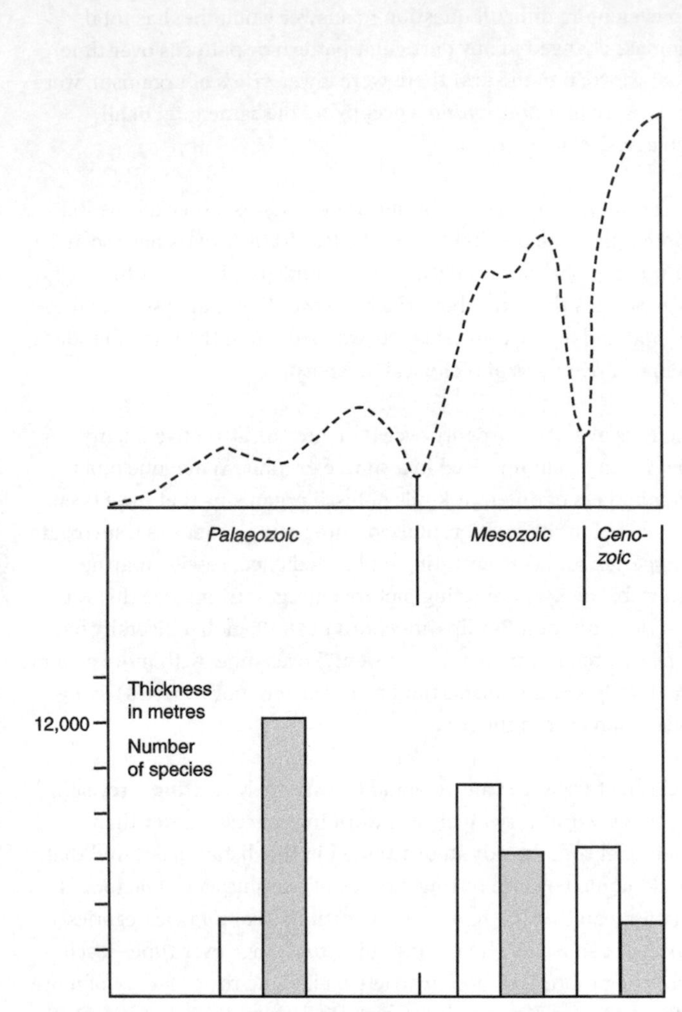

18. Changing numerical diversity of life over time, as first estimated by Professor John Phillips at Oxford in 1860. Above, his compromise view; below that, his estimates for the thickness of rock in metres (shaded) and numbers of marine fossil species (unshaded), from which he calculated species-per-hundred-metres ratios of 201 for the Cenozoic, 150 for the Mesozoic, and 26 for the Palaeozoic

emerged – mammals or insects – for example, has occupied an ecological niche that had previously been empty or only partially colonized. Or that something chemical (perhaps the amount of atmospheric oxygen or carbon dioxide, oceanic iron, or calcium) has changed globally. Ocean and atmospheric circulations may have changed, probably in connection with plate tectonics. Indeed, all of these factors have probably applied at some time in the past.

Phillips and those following him realized that a number of factors could significantly bias the raw data. His first tallies actually showed a peak of diversity in the Mesozoic, undoubtedly because the Mesozoic is far more heavily represented in the rocks of Great Britain than is the Palaeozoic or Cenozoic. So he recalibrated his data according the thickness of the available strata from which fossils had been collected. Later, it became obvious to qualify the numbers according to the exposed surface area of strata as well.

It was self-evident that a major bias will necessarily result from the fact that we will have a better record from the more recent strata, both because the rocks are better preserved and we also know modern animals and plants better than older ones. This is called 'the pull of the recent' and must inevitably contribute to the impression that there are more species alive now than ever in the past. Also, fashions vary, with some parts of the record dominated by palaeontologists who are 'splitters', naming new species on the basis of very small differences. Others may be the province of 'lumpers', doing just the opposite. This is particularly annoying for students learning about human evolution; for example, for one author *Homo erectus* is a single species, for another it is three or four.

There is a danger of circular argument when it comes to stratigraphic boundaries, most of which were originally defined by the presence or absence of particular species: if those species later turn up on the 'wrong' side of the boundary they will be given different names. Other scholars have assumed that faunas from geographically distant faunal or floral provinces must be different, so workers on different continents describe new species for entities that have already been named. There is a special irony here when it turns out that, because of continental drift, regions such as Maritime Canada and North-Western Europe were contiguous in the Palaeozoic. There is even an 'imperial' element to this – workers from one country will name new species from abroad according to what they know at home. The British in a particular foreign country would identify species differently from the French or Germans. Happily, most such practices are behind us, but that still leaves the problem of clearing up what was written in the past.

Putting all this together, in the 1980s four leaders in the field (David Raup, Jack Sepkowski, Richard Bambach, and James Valentine) converged on the view that the data genuinely show an increase in taxonomic diversity on earth over time, although the increase was far less dramatic than originally thought. But today even that compromise position seems shaky. It is apparent that the existing raw data need major revision, which will be a huge task

Geology and evolution

The study of the earth's interior may conduct us, if not to a solution of the great problem of Creation, at least to a knowledge of some of the laws by which it was governed at different epochs. It has cast much light on the point. It shows us that organic beings became more and more perfect from the commencement of life on earth to the time of man's appearance. It shows us that during the long interval separating man from the first arrival, the universe was agitated by successive revolutions; but that since that time the equilibrium has become perfectly established, so as to permit man to spread across the globe.

M. Rozet, *Traité Elementaire de Geologie* (1835)

given the number of described species and the number of specimens in existing collections. Work is now under way to refine the raw data, and new hard-nosed analyses already suggest something entirely more logical, if less dramatic, than an ever-increasing diversity. It seems to be the case, after all, that the earth achieved maximum taxonomic diversity during the mid-Palaeozoic and that overall diversity has been essentially stable (with considerable fluctuation around a mean) ever since.

Extinction: the world of the Red Queen

Extinction is the one universal in all of palaeontology. All species live in a changing environment in which, as the Chicago-based palaeontologist and evolutionary theorist Leigh van Valen has pointed out (paradoxically drawing the right conclusion from a flawed analysis), the Red Queen from Lewis Carroll's *Through the Looking Glass* rules: 'A slow sort of country', said the Queen. 'Now, HERE, you see, it takes all the running YOU can do, to stay in one place. If you want to get somewhere else, you must run at least twice as fast as that.' In nature, species evolve to meet environmental needs, but eventually always fall behind, being replaced by new species that eventually succumb in their turn. Most fossil species survive only for 2 to 4 myr and genera for 5 to 20 myr (long-lived exceptions include bivalve molluscs, reef corals, and planktonic foraminiferans).

Beyond the constant turnover of species ('background extinction'), all tallies of Phanerozoic fossil diversity reveal a series of pronounced drops (six or seven in number) in total global diversity, spread over time. Analyses show a huge disparity between the rates of extinction and origination during these relatively brief periods of 'mass extinction'. Because of the extinction of dinosaurs, for the public the best-known loss of diversity came at the end of the Cretaceous. Going further back in time, there was a smaller extinction towards the end of the Triassic. Numerically, the greatest extinction event of all was at the end of the Permian. There were at least three Palaeozoic mass extinctions: in the Late Devonian, Late Ordovician, and Late Cambrian. By adding in smaller level extinctions (for example, in the Late Ordovician and Late Eocene), some authors have even attempted to show that extinctions occur with a 26-million-year periodicity.

Perhaps the most important thing to be said about these episodes of mass extinction is that they were not universal. At the end of the Cretaceous, land vertebrates as apparently frail and vulnerable as salamanders, birds, and mammals sailed through. Crocodiles and turtles survived while the other large reptiles, many land plants, and all the ammonites did not. The effect was greatest in the marine realm, as it was also in the Permian extinction. The Cretaceous extinction is popularly described as having killed off the dinosaurs but only killed the last few different kinds; dinosaurs as a group had been waning fast throughout the latter part of the Cretaceous. This may be true of other groups and other extinctions. As for 'suddenness', while the Cretaceous extinction is thought to have occurred over a few thousands of years at the very most, the Permian event may have lasted for 1 million or more.

There is an outside possibility that the scale of these extinctions has been grossly exaggerated by biases in the record. If, on the other hand, these mass extinctions are real, they must have been caused by forces acting outside (or at the far extreme range of) typical uniformitarian processes, and over a timescale measurable in ecological rather than geological time. It is unlikely that all had the same cause. The Cretaceous and Permian extinctions, for example, were accompanied by mass marine regression and extremely high volcanic activity that spread huge sheets of lava over vast distances (the Deccan and Siberian traps, respectively). Combinations of climate changes (high and low temperatures), atmospheric fluctuations in oxygen and carbon dioxide, sea level changes, and changes in habitat diversity resulting from continental re-arrangements and mass volcanism are still high on most people's lists for the causes (whole or partial) of mass extinctions.

Asteroid impact has attracted the most publicity and is the current popular causal explanation for the Late Cretaceous extinction. The prime evidence is the presence of unusual elements such as iridium in the affected depositional environment. A possible site is the Chicxulub impact crater in Mexico. Presumably such an event would have been followed by extensive forest fires and the deposition of large amounts of fossil carbon, but the evidence for that is equivocal. There is even doubt about the precise synchronicity of the impact itself and the extinction events.

The tempi and modes of evolution

Before 1944, the field of palaeontology was dominated by relatively old-fashioned morphological, taxonomic and stratigraphic studies. Then George Gaylord Simpson published his path-breaking book *Tempo and Mode in Evolution*, in which he used quantitative analyses and statistical methods to probe deeper into the fossil record, and in particular, to bring the study of fossils on a par with the study of living organisms.

No analysis is better than the raw data allow, and (as noted above) there is good reason to be cautious about a lot of the numbers currently used in palaeontology. However, Simpson showed palaeontologists that their data were in places complete enough to yield information on variation, one of the cornerstones of Darwinian natural selection. He produced the first quantitative estimates of the rates of evolution. He derived 'taxonomic rates of evolution' by measuring the longevity of different species, genera or families or, most simply, by plotting the distribution of first and last appearances, over a suitably fine timescale. These can be expressed as rates of either origination and of extinction and the evidence shows that these clearly vary within and among lineages and/or at certain time periods.

Rates of morphological evolution within a lineage can be estimated by scoring fossils for particular anatomical features such as the proportions of a tooth or the number of separate bones in the skull roof. This work grows directly out of the work of d'Arcy Thompson and Julian Huxley on the mathematics of relative growth, and in turn leads to analyses of differential growth rates (heterochrony) and other developmental phenomena in fossil organisms. One of the classic cases where palaeontology has documented evolutionary change is in the evolution of horses, dog-sized five-toed ancestors in the Eocene giving rise to radiations of three-toed and then single-toed species. In the process of becoming larger, horses have also developed disproportionately longer limbs (for running) and longer heads and taller teeth (for chewing grasses). These changes proceeded at different rates in different lineages within the horse family.

Simpson showed that different groups vary, evolve, and become extinct at different rates and that those rates could be measured. His approach lives on even though his data and many of his arguments have long since been eclipsed by the extraordinary growth in quantitative palaeontological science of the last 50 years. That was only to be expected but, curiously, some palaeontologists have recently complained that he had demoted their subject to the status of a mere handmaiden of biology. In this, they forget where the subject had been before Simpson, to whom a huge debt is owed for showing that palaeontology can often reveal phenomena that are invisible in ecological time, but on the same terms. However, it may also be true that Simpson, who dared to call his book as fusion of palaeontology and genetics, was more optimistic about the adequacy of the fossil record than we are today.

The shape of evolution

It always used to be thought that the best fossil record should reveal slow and gradual evolutionary change, one species morphing into a slightly different one or slowly dividing into two. Darwinian transmutation of species should be observable simply by climbing far up the face of an exposure, collecting specimens inch-by-inch, year-by-year, as you go. However, in 1977, Nils Eldredge (at the American Museum of Natural History) and Steven Jay Gould (at Harvard) found a different pattern. They documented cases where species remained without apparently evolving (in *stasis*), for long periods of time in the record and then were quite suddenly replaced by closely related (apparently daughter) species. They called this pattern 'punctuated equilibrium', to distinguish it from evolutionary gradualism – seeing the shape of evolution as being more of a staircase than an incline.

Punctuated equilibrium was a literally revolutionary concept and could be construed (incorrectly) as hinting at an old-fashioned saltationism (the idea that evolution proceeds by discordant leaps) and (correctly) revealing a touch of Marxism. In fact, punctuated equilibrium fits well with the model of speciation that involves the isolation within a species of peripheral populations in which change is rapid, followed by reinvasion and replacement in the parent territory where a more stabilising selection has been in force. Such models have been popular because of the difficulty otherwise in explaining why new variants would not be swamped out by interbreeding in the main populations (a problem that Darwin partially fudged by adopting a modified Lamarckism in the fifth edition of *On the Origin of Species*). One interesting aspect of punctuated equilibrium is that, if evolutionary change is concentrated in rapid speciation events rather than accumulating during a long gradual history of any species, then different taxonomic rates of evolution are explained, not as different modes of change, but as the result of different durations of the periods of stasis.

Ad hoc explanations of all this abound. Evidence from cases where several contemporary species seem to undergo 'punctuation' simultaneously suggests that the speciation events may be driven by relatively abrupt (rather, again, than gradual) external environmental events. Elisabeth Vrba at Yale University calls this the 'turnover pulse hypothesis'. In a developmental explanation, a set of populations subject to constant environmental stress would be expected to build up a number of small genetic changes (invisible in the fossil record). Once these built up to a critical mass, a threshold event in the genetic control of development would occur, leading suddenly to expressed change (the phenotype).

Testing any of this brings us up against the sampling problem. Logically, one would expect any species to be represented by a number of populations encompassing a range of intra-specific diversity. If sampling in a single particular place shows gradual change, it might simply be that the environmental conditions there have slowly shifted and a different sub-set of the existing populations had moved into the sampled ecological space. If sampling shows no change, it could be that the characteristics of the main body of the species may have shifted, but elsewhere, while the sampled deposits merely reveals an outlier population that has remained static.

Once the concept of punctuated equilibrium had been launched, as is often the case, a bitter, personal debate ensued between traditionalists and revolutionaries (the gradualist 'creeps' and the punctuational 'jerks'). In fact, it is hard to see why one group of palaeontologists would be custodians of the 'truth' or that only one mode of speciation would hold true for the whole fossil record. Perhaps the greatest benefit of the debate was to launch two important efforts: a rigorous analysis of previous examples that had thought to demonstrate gradualism, and a search for geological settings in which a really fine scale of evolution over time was demonstrable in a single, uninterrupted, time span. The only way to approach the problem, and *then only case-by-case*, would be to sample the entire geographical distribution of all the species in question, over a year-by-year time frame in order to analyse both temporal and spatial diversity of all the included populations during a speciation event. Incontrovertible evidence for gradual or punctuated speciation turns out to be sparse because of the problems of sampling so extensively. The present consensus view is that different models – gradual, punctuated, combined – hold true for different times, different places, and different organisms.

There is an irony here. Most early philosophers had argued that species are fixed and unchanging in nature; in Judaeo-Christian philosophy, for example, they were fixed at the moment of Creation. Alternatively, in both the transcendental progressivist and the gradualist evolutionary viewpoint, species may be an artefact. What we see as a species in ecological time may be simply a snapshot taken today of an entity in constant flux. Most early evolutionists argued fiercely for the concept of gradual transmutation of species. Punctuated equilibrium, on the other hand, returns us to a revised concept of the fixity of species worthy of the American baseball hero and wit, Yogi Berra: species don't change until they change! And it requires that the traditional concept of species transformation – gradual change that, in principle, should be observable in the fossil record – be replaced by a scheme in which it would not.

Major evolutionary transitions and the macro-evolutionary problem

The origin of a major new group of organisms is defined by (and appears in the fossil record as) novel structures and the occupation (take-over) of a radically new environmental niche. Obvious examples would be the origin of land vertebrates (with legs and lungs) from fishes, and of both birds and mammals from (different) reptile stocks. But such events are usually among the most poorly documented areas of the record. And that leads one to ask whether the relative scarcity of transitional fossils reflects something special about the processes involved, or whether the problem is simply another artefact of the fossil record.

In the origin of major groups, evolution seems to have proceeded unusually quickly and at low diversity. This raises the question of the minimum number of speciation events required for any of the major adaptive shifts in evolution that result in the origin of new major groups (like birds, mammals, insects, or the flowering plants). Previously, a major evolutionary shift, say from reptile to birds, had to be explained by piling up hundreds and thousands of such events and huge numbers of species each differing only slightly from each other. But this 'accumulation' interpretation of evolution is either false or insufficiently true. Many phases of evolutionary change appear to involve change at a rate that cannot be accounted for by slow gradual evolution. In the origins of major groups, evolution appears to be very fast and transitional forms were less diverse and less numerous than in either the ancestral or descendant groups. Not until the new adaptive structure and physiology had been tried and tested did broad diversification begin again.

This suggests a major uncoupling of morphological (change in structure) and taxic (numbers of species) modes of evolution or, at least, a high rate of speciation and morphological change within a narrow lineage without lateral diversification. The simplest explanation may be that all this is an artefact of sampling from the transitional environments. For example, in the reptile to bird transition, the transitional forms may have occupied a woodland environment from which preservation is poor. Perhaps, also, they had been geographically confined to a small area. The apparent speed of change might then simply be due to gaps in the record. But it seems unlikely that this explanation would apply in all cases of major evolutionary transition.

Darwin himself thought that transitions occurred with low diversity at the individual level: 'Intermediate varieties, from existing in lesser numbers than the forms which they connect, will generally be beaten out and exterminated during the course of further modification and improvement'. In which case, fossils will be scarce in the transitional process, but that does not explain the apparent high speed of morphological change.

The concept of 'key innovation' may help here. A key innovation is a morphological, physiological, or developmental change that might itself be minor in effect but opens up a wide range of new possibilities. It arises innocently enough in one context and then turns out to be even more useful in a different context. The feather is often thought of as such a key innovation, evolved first for temperature control. The tetrapod limb arose out of modifications of fish fins for living in shallow water. The lung arose from an accessory respiratory device and/or buoyancy device. And so on. In such cases, a lot of the hard work had been done before the actual transition.

A second concept is 'correlated progression' (named by this author with a nod to Darwin's concept of 'correlated variation' – an example would be the notion that all blue-eyed, white cats are deaf). No part of an organism exists independently of the rest. The first fish that held an air bubble in its gut simultaneously created new possibilities in respiratory gas exchange, in buoyancy control, and in underwater hearing. Normally, the potential for change in particular organ and physiological systems, and their developmental pathways, is constrained because of the other systems with which they are bound up, functionally and developmentally. In isolation, each could change just so far and no further. In correlated progression it is postulated that, if they were to change together, the potential for innovation would be greater. For example, in the vertebrate head there are elements in common among the jaw mechanics, respiratory mechanics, postural mechanisms, and hearing mechanisms. Change in any one would produce a small effect. Changes in several together, under the same adaptive regime, would become self-reinforcing and produce a much larger effect.

Both 'key innovation' and 'correlated progression' make more sense when translated to a developmental rather than adult context. For example, in the evolution of horses, the three-toed condition arose in parallel more than once in the Miocene and the one-toed condition more than once in the Pleistocene – suggesting that the driving force was not just selection on adult phenotypes but some significant long-term shift in developmental pathways that produced the variants on which selection acted.

One attraction of the punctuated equilibrium model is that it might allow the potential for speciation to occur faster than under the gradual model (by shortening the periods of stasis). It also points us to the realization that selection must act at more than one level. In a fully hierarchical view of evolution, just as there is selection at the gene, individual, and population levels, there is also selection at the species level and possibly even selection at higher levels. In this case, the difference between, say, the bushy pattern of evolution seen in many groups, producing dozens of extremely similar species (most insects, humming birds) and the low diversity of a group in transition may in part be due to degrees of selection acting at the species level and driven by the rigours of environmental conditions. As the group rapidly becomes more adept (morphologically, physiologically, and behaviourally) in dealing with a range of the new conditions, species selection eases and diversification increases.

Living fossils

Darwin's theory of natural selection put the whole subject of evolution squarely on the scientific map and created a new range of expectations for the power of palaeontology. But Darwin was aware that the fossil record did not provide as conclusive evidence for his theory as he might have liked; there were too many gaps due to the imperfection of the geological record. There were also inconsistencies in the record of living organisms. In his *On the Origin of Species*, he articulated the concept of the 'living fossil'. Living fossils (an oxymoron, since living organisms cannot be fossils) are the exceptions that prove the evolutionary rule; they seem more immune to the pressures of change over time and relentless extinction than other groups. Typically there survives just one (or a few) species from a group that was previously more widespread in the ancient fossil record.

Living fossils are the last remnants of lineages that evolved at dramatically slower rates than the norm. They are survivors from ancient nodes along the tree of organic diversity, displaying the structure (and presumably the biology, physiology, and chemistry) of an ancient time. They are often of enormous value to science, especially when it comes to reconstructing the life of related fossil species. Altogether, we can list some 30 to 40 living species that all fall into this category. Among the examples that Darwin knew were several kinds of tropical freshwater fishes, including three genera of lungfishes (Dipnoi), the African bichir (*Polypterus*), and a remarkable egg-laying mammal, the duck-billed Platypus (*Ornithorhynchus*). He thought that other examples would be found in the deep sea.

Perhaps the best-known living fossil is the coelacanth *Latimeria chalumnae*, a fish first discovered alive in the Indian Ocean (although not at great depths) in 1938. It belongs to a group first known as fossils in the Devonian but thought to have become extinct with the dinosaurs. When a modern *Latimeria* and a Devonian *Nesides* are compared, the differences, in the skeleton at least, are staggeringly few. In fact, coelacanths should have been interesting long before *Latimeria* was found, because it was already obvious from comparison of Late Cretaceous and Devonian fossil coelacanths how very little they had changed. Discovery of the living species added many new possibilities for study and evolutionary biologists naturally infer that conservatism in skeletal evolution might indicate a similar conservatism in the rest of the fish's biology. In studying a living coelacanth, we hope we study a living Devonian fish. Studying the biomechanics of the fresh coelacanth that I obtained at Yale University in 1966 was one of the great highlights of my scientific career.

The concept of a 'living fossil' is confusing in that it is the *lineage*, not the species, that survives in spectacularly conservative fashion. We have no evidence to suggest, for example, that the species *Latimeria chalumnae* has survived for more than the usual few million years; the genus *Latimeria* is unknown as a fossil. There are very few individual species that seem truly to have survived in the fossil record – as that same species – for longer than the usual allotted span.

What all so-called living fossils have in common is that they – as species or as representatives of a lineage – demonstrate an extremely slow rate of structural evolution, accompanied by relatively low species diversity over time. Meanwhile, close relatives of those same groups evolved normally, species-by-species. A distant cousin of the Devonian coelacanths and lungfishes, for example, was our own ancestor (and that of all land tetrapods).

No-one knows why and how living fossil lineages have persisted. Darwin thought that the pressures of selection in the ancient freshwater lakes in Africa were, for some reason, less: 'They have endured to the present day, from having inhabited a confined area, and from having thus been exposed to less severe competition.' But that is hardly a full answer. By no means all living fossils have a restricted geographical distribution; the horseshoe crab (remnant of a Palaeozoic lineage related to spiders) is found worldwide. Some living fossils may have a 'generalized' and adaptable (rather than highly specialized) lifestyle that allows them to cope with environmental changes that wiped out their contemporaries. Some may have survived through the opposite strategy – being adapted to a particular quite specialized mode of life, the niche for which has survived unchanged over time, or for which a successive series of successive niche changes have been adaptively neutral. Possibly the slower rates of evolution in living fossils are, after all, just the result of chance. Perhaps one should also reverse the question and ask how, and why, evolution proceeds so quickly elsewhere!

Missing links

Ever since Aristotle, the living world has been seen as some version or other of a continuous 'Chain of Being'. Evolutionary theory gave this a causal framework that confirmed the continuity of life in space and time and turned it into a 'Chain of Becoming'. Our genealogical record from fossils is, however, full of gaps. These gaps are interesting because we know where they are, and what, in general ought to be filling them. We term these undiscovered species 'missing links'. A missing link is that species or sequence of species that is missing from our database because of gaps in the record, but must exist in principle. The feathered reptile/toothed bird *Archaeopteryx* – a perfect 'missing link' – was, by happy chance, discovered in 1860, giving a new reality to burgeoning world of evolutionary theory. Over the years, a great deal of palaeontological effort has been devoted to finding other such missing links.

Strictly, 'missing link' is a metaphor. Originally, it was used as a predictive term, referring to something that has not yet been found. In that sense, it is an hypothetical construct. Today, the term is most frequently used to describe the discovery itself and in that sense represents confirmation of an hypothesis. In this latter, most widely used, sense '*missing link*' is now another oxymoron, referring to the discovery of an organism (therefore no longer missing), usually a fossil, that occupies an intermediate position in the record of two otherwise separate lineages.

Largely through the writings and lecturing of Thomas Henry Huxley, the great anatomist, palaeontologist, and Charles Darwin's staunchest supporter, the term 'missing link' has always had a special significance in the world of human palaeontology where missing fossils link us to the great apes. Because of this, unfortunately, the term also became an item of personal abuse early on, Victorian humorists often finding it witty to describe the Irish labourers at the time building English roads and railways as the 'missing link in human evolution'.

There is often considerable overlap here with the term 'living fossil'. For example, when discovered in 1836, the South American lungfish *Lepidosiren* was thought to be the previously missing link between fishes and tetrapods – not surprisingly, since it has lobed fins, lungs, and a sort of internal nostril or choana. The egg-laying monotreme *Ornithorhynchus* similarly connects reptiles and mammals. (Unfortunately, the living fossil *Latimeria chalumnae* turns out only to be another coelacanth and not a direct link to tetrapod origins.)

Chapter 8
Of molecules and man

One late August day in 1963, I perched part way up a small cliff near the west bank of Lake Turkana (then still called Lake Rudolf) in northern Kenya and looked with excitement and awe at the fossil human molar tooth that I had just picked up. How much like me had the owner of that tooth been? How had that person been related to the present Turkana people of the region? How long ago had he or she lived? I felt once again that curious mixture of feeling both 'the strange' and 'the familiar' attributes of fossils. Like the dinosaurs that we (especially as children) see as half-real, half-unreal, human fossils are 'of us' and 'not of us' at the same time; we marvel at the similarities to ourselves and are intrigued by the differences – especially in those features that the Victorians used to portray our ancestors as shambling brutes.

No fossils fascinate us more than human fossils, and new discoveries crop up today at a surprising rate, giving special meaning to Alexander Pope's aphorism: 'The proper study of mankind is man.' Human fossils give us a whole extra perspective about who we are and how we came to be. They show us how our forebears lived, where they lived, how they moved, what sorts of food they ate, how big their brains were, perhaps even whether they could speak or not. The fossils show, at least in outline form, our transformation from quadruped to biped, herbivore to omnivore, and give us hints as to the development of social structures, intelligence, and culture.

The human fossil record also affords us a telling case study in all the opportunities and problems of palaeontological science. It is salutary, if also frustrating, to write about an area of science in which one knows that the raw data continue to change yearly, even monthly. On the other hand, because the human record deals with material of relatively recent age with a decent chance of preservation, and the skeleton comprises such a huge number of measurable characteristics, from cranial capacity to the finest details of the teeth or limbs, we can analyse evolution acting at a finer scale and over a shorter time span (at a scale of tens of thousands of years) than in most areas of the fossil record. As a result, the human record offers a decent chance that we will eventually be able to fit together a more finely detailed reconstruction of a fossil genealogy than for almost another other group.

And here, where palaeontology meets archaeology, is also one of the places where modern molecular science can be used side by side with the more traditional kinds of information from fossils, though the findings are not always in agreement. A few human fossil remains are sufficiently recent in age (less than 100,000 years old) that we can extract parts of their DNA. Using simple assumptions about the rates at which mutations are fixed in the RNA and DNA molecules that encode our very existence, it is possible not only to work out which species are most closely related to each other, but also to estimate how long ago the lines leading to them diverged. Molecular analyses not only tell us when modern humans arose, but can provide answers to the old question of whether they (we) simply wiped out all the other species (like the Neanderthals, with the last of whom they lived more or less side by side) or assimilated them.

Until the mid-19th century, it was both an empirical fact that there were no known human fossils and a matter of principle that there could be none. The longest-surviving element of Judaeo-Christian belief concerning Creation had been the transcendent status of man. Over the last two centuries, as the fossil record made it increasingly necessary to yield the point that faunas and floras have changed over time, two key points were defended to the last: that God was responsible at least for the natural laws that control all creation (even if it did not occur in a single event) and that man had been God's direct creation, independent of other events, processes, and causes in nature.

The discovery of human fossils in Neander Valley of Germany in 1856 changed all that and also coincided with the advent of Darwinian theory, within which humans are subject to the same natural laws as all else in this world. The next great discovery came in 1891, when Eugene Dubois, under the influence of Humboldt, set off to find fossil man in Indonesia and succeeded brilliantly with the discovery of Java Man – *Pithecanthropus* (now *Homo*) *erectus* – in every sense a missing link between man and the apes. The subsequent discovery of European Cro-Magnon man seemed to bridge the gap between the Neanderthals and modern humans. Then in the 20th century, the focus of attention moved to Africa and to a cascade of discoveries about the earlier phases of our ancestry.

In 1921, the Broken Hill (Zambia) skull was described. Three years later, Raymond Dart in South Africa discovered the first *Australopithecus*. Robert Broom's discoveries in the 1930s and 1940 firmly placed *Australopithecus* and *Paranthropus* at the root of human ancestry. After the war came a major period of discoveries in East Africa at Olduvai Gorge and other sites, by L. S. B. Leakey and his wife Mary, and then later their son Richard. In the 1970s, Donald Johanson and others made magnificent discoveries in Ethiopia. In the last 30 years, the history of early hominids has been substantially fleshed out with an explosion of breathtaking discoveries throughout Africa, Europe, and Asia. There is no reason to believe that this will not continue for a long time.

Even before the first human fossils had been discovered, a growing body of evidence had already shown that human history would one day be traced back into the tangle of primate relationships represented by the living apes, monkeys, and their relatives. Central to this was the discovery that the living great apes – the orang-utan (discovered in 1778), the chimpanzee (1788), and the gorilla (1847) – are so very similar to us anatomically and behaviourally. When chimps and orang-utans were put on display at zoos, dressed up as children to hold tea parties, the effect on the Victorian public was staggering. From then on, whereas a relationship between a man and something like a monkey might be far-fetched, as Thomas Henry Huxley put it, metaphorically at least: 'I would not be ashamed to have an ape for a grandfather.'

The term 'human' is variously taken to mean just our own species, *Homo sapiens*, or the members of the family Hominidae in which we belong, together with all our extinct relatives, the closest of which are in 'our' genus: the fossils *H. neanderthalensis*, *H. heidelbergensis*, *H. erectus*, *H. ergaster*, *H. rudolfensis*, and *H. habilis*. The species *H. rhodesiensis*, *H. antecessor*, and *H. mauretanicus*, and a few others are less uniformly agreed upon.

Darwin at the Zoo

She (Jenny the orang-utan) threw herself on her back, kicked & cried, precisely like a naughty child. – She then looked very sulky and after two or three fits of passion, the keeper said, 'Jenny, if you will stop bawling & be a good girl I will give you the apple'. – She certainly understood every word of this, &, though, like a child, she had great work to stop whining, she at last succeeded, & then got the apple, with which she jumped into an arm chair & began eating it, with the most contented countenance imaginable.

Charles Darwin, letter to his sister Susan Darwin, 1 April 1838

The genus *Australopithecus* includes *A. africanus*, *A. anamensis*, *A. ramidus*, and *A. afarensis*. *Paranthropus* includes *P. boisei*, *P. aethiopicus*, and *P. robustus*. All these lived in the last 5 million years. The Hominidae plus the apes (Pongidae) make up the Hominoidea. Thus all humans are hominids, but not all hominoids are human.

Analyses of the DNA of hominoids confirm that the closest relatives of humans are the chimpanzees. While we may look different facially from chimps as adults, the resemblance between a baby chimp and a human amply confirms the molecular evidence. We are not descended from any kind of actual chimpanzee, however; instead, the molecular evidence shows that the lines leading to modern (and extinct) humans on the one hand, and to modern chimpanzees on the other, diverged some 6 million years ago. Three relatively recent discoveries of very primitive hominids – *Ardipithecus* from Ethiopia (*A. kadabba* dated at 4.2 myr and *A. ramidus* at 4.5 myr), *Sahelanthropus* from Chad (between 6 and 7 myr), and *Orrorin* from Kenya (6 myr) – seem to confirm this. Morphologically, they appear close to the human–ape transition. They are either precursors of modern apes, or of humans, or of both.

At the modern end of the family tree, a number of molecular studies have given dates for the oldest common ancestor of all living humans at between 400,000 and 120,000 years, with the latest studies suggesting a date of 175,000 plus or minus 50,000. As these dates are based on mitochondrial DNA, which is only inherited maternally through the egg, this oldest point of divergence has been dubbed 'Eve'.

DNA evidence cannot reach very far back in time. Only fossils can tell us what happened between the chimp–human divergence and the origin of thoroughly modern humans. So far, the period between 5 myr and about 2.5 myr ago can be shown to have been dominated first by *Australopithecus* and then by *Paranthropus* (living about 2.7 myr to 1.3 myr ago). Neither of these genera has been found outside of Africa. The earliest known *Australopithecus* is *A. anamensis* (4.2 myr) from Kanapoi and Allia Bay in Kenya. *Australopithecus afarensis* – especially as represented by the famous 'Lucy' skeleton from Haddar, Ethiopia. Lucy, discovered by a group led by Donald Johanson in 1974, had a primitive kind of bipedal locomotion. Not only do the structure of the hip and knee joints indicate an upright posture, track-ways from Laetoli, Tanzania, definitely show a bipedal animal with a version of a striding gait. These were among the first hominids to make the transition from forests to the open woodlands that were then extending rapidly across the African landscape, driven by a phase of climatic cooling. Their brains (around 400–500 cm^3) were still relatively small compared with *Homo*, but were quite large in comparison with a chimpanzee. The dentition suggests a diet in transition from a wholly vegetarian diet of fruits, roots, and leaves to a mixed diet including small animal prey. They lived in extended family groups.

From something like *Australopithecus*, two lines of evolution emerged. *Paranthropus* was a sideline of hominids in which the body build was heavy and 'robust'. They became extinct by about 1.4 myr ago, their last representative (so far) being Leakey's famous *Paranthropus* (formerly *Zinjanthropus*) *boisei*, which acquired the name 'nutcracker man' from its huge lower jaw and teeth. The second line led to the genus *Homo*.

Our immediate ancestors

Over the past 30 years, a general picture of the immediate ancestry of *Homo sapiens* has become clear, although there is still much to be done. The subject offers a good example of the difficulties that apply when one looks into the finer details of any fossil lineage(s). Undoubtedly there are still gaps in the story. Because older discoveries tended first to be assigned to new species and even genera, a great deal of work has had to be done in clearing up old taxonomies and the field is littered with names such as *Pithecanthropus* (du Bois's 'Java man') and *Sinanthropus* ('Peking man') that are no longer thought valid. Study of human fossils is further complicated by the (invaluable) fact that, almost monthly, the 'earliest date' for any piece of evidence is pushed back. Less than 10 years ago, for example, the date of human migration out of Africa was 1 myr ago; now it is 2 myr ago. As more and more fossils are found, to hijack a quote from my old teacher Alfred Sherwood Romer on mammal origins, 'increasing knowledge leads to triumphant loss of clarity'.

One of the more difficult tasks is to fix the distinctions among species. This looks easy when you have species separated by tens of millions of years, but at the fine scale of resolution in the fossil record where distributions overlap and the structural features seem to grade one into another, it is much harder. Variation at the population, species, or even generic level turns out to be difficult to calibrate. However, setting all hominid fossils apart from the (other) great apes are: reduction in the size of the dentition, large brain capacity, and a more domed cranial vault, various aspects of the skeleton to do with a bipedal upright stance, and, in the later stages, some kind of culture involving crudely made tools.

Currently the oldest members of a broadly defined genus *Homo* are *H. rudolfensis*, from about 2.5 myr old localities east of Lake Turkana, Kenya, and *H. habilis*. These were followed quite quickly by *H. ergaster* and *H. erectus* (du Bois's original *Pithecanthropus*).

How to distinguish species

Tattersall's law states that if you can tell two skulls apart at fifty paces, you have two genera, while if you have to scrutinize them up close to tell the difference, all you have is two species. Of course, this is an oversimplification . . .

Ian Tattersall, *The Fossil Trail* (1995)

Some workers divide an Asian species *H. erectus* from separate African species *H. antecessor* and/or *H. mauretanicus*. *H. rudolfensis* had a cranial capacity of 700–800 cm^3; that of *H. habilis* was smaller at 500–700 cm^3, *H. ergaster* was 'brainier' still at 600–1000 cm^3, and *H. erectus* had a cranial capacity of 900–1200 cm^3.

Somewhere between 1 myr and 800,000 years ago, the ancestors of *Homo sapiens* arose possibly out of *Homo ergaster* via *Homo antecessor* or (in a rival scenario, see below) from *Homo erectus*. The brain in modern *Homo sapiens* ranges from 1200 to 1800 cm^3, averaging at 1400 cm^3. But, when you include all the fossils, *Homo sapiens* itself turns out to be difficult to define. The brain is very large, the cranium high and rounded, the face vertical: but *how* big, rounded, or vertical? In contradiction to the modern fashion elsewhere in palaeontology for 'punctuated equilibrium' as a model for evolutionary change, workers with human material tend to see things gradualistically and several kinds of *Homo sapiens* are currently recognized. An archaic form seems to have existed from about 250 thousand years (kyr) ago in the form of inconclusive material fossils from a range of African sites. The best-preserved 'early modern' *H. sapiens* come from 160-kyr-old specimens from the Herto locality in Ethiopia; related material from Kenya is dated at 195 kyr. This fits pretty well with the molecular 'Eve' data.

The oldest thoroughly modern *H. sapiens* come from the Klasies River Mouth caves in South Africa and Qafzeh cave in Israel, dated at some 90 kyr and 115 kyr, respectively. Advanced features included a large cranium, globular braincase, and a domed forehead, but even so they retained a more archaic, wide inter-orbital breadth, low nose and flat mid-face than in present populations. *Homo sapiens* seems to be a work in progress.

It is even more difficult to find morphological evidence for the origins of complex cultures and, by implication, the first evolution of a conscious, rational intellect. (Something like the origin of

19. The oldest thoroughly modern humans (around 100,000 years old) were found at Qafzeh cave in Israel

speech is particularly hard to pinpoint, because the key cartilagenous elements of the larynx are not preserved, although one can make inferences from the thorax and skull base.) Nonetheless, there are some milestones. Tool-making has usually (as a matter of principle) been seen as a unique characteristic of the genus *Homo*. In fact, tool-making of the most primitive 'Oldowan' type seems to have started around 2.5 myr ago, either in late *Australopithecines* or in *Homo ergaster*, and somewhat predates the development of a significantly larger brain, which occurred somewhere in the *Homo habilis*/*Homo ergaster* divergence. These tools consisted of simple hand axes and sharp flakes of stone used as knives and scrapers – the Stone Age had begun. Presumably these early people had much earlier used sharpened sticks, possibly with their points hardened in fire, as tools, but the earliest fossil evidence for controlled fire dates back to some 790,000 years ago (*H. erectus* or *H. ergaster*).

Around 1.5 myr ago, *Homo ergaster* made a further significant innovation, one that in many senses prefigures the entire history of human technology. In the Acheulean tool type ('the Swiss Army Knife of the Palaeolithic'), flakes were removed from both sides of a stone core (quartz, flint, obsidian) so as to shape a complex tool – a more sophisticated axe that could puncture as well as crush, a knife, and even eventually a spear head or an arrow head. What kind of intelligence, social structure, or culture these large-brained early humans enjoyed is not possible to tell. How did they communicate: by grunts and gestures, or something more advanced? Cave painting ranges back only to about 35,000 BC. The making of beads out of shells has recently been suggested to range back to 70,000 BC. Therefore, there must have been a very large gap in time between the evolution of what we see as a thoroughly modern brain (and advanced skeletal anatomy) and the inception of modern patterns of use of that brain.

A word needs to be added here about Neanderthals, which popularly still conjure up images of brutish, shambling ape-men. In fact, Neanderthals were big people, intelligent hunters with a short, powerful physique; a big, long head with very heavy brow ridges; a large nose; and a weak chin. They lived in family groups with a quite small home territory. It used to be thought that *H. neanderthalensis* was the ancestor of *H. sapiens* – things were easier when so few fossils were known! Most current opinion (but not unanimously) holds that *H. neanderthalensis* arose independently from within *H. heidelbergensis* or from a separate stock within *H. erectus* or *H. rhodesiensis*. The palaeontological evidence shows that in the Levant, around 100,000 years ago, *H. sapiens* and *H. neanderthalensis* lived in the same region and possibly, sometimes, together. They are both associated with advanced stone tools and with ritual burials. In Europe, the story may have been different – the arrival of *H. sapiens* (no more than 50,000 years ago) being followed relatively quickly by the extinction of Neanderthals (at about 30,000 years ago). The obvious possibility is that *H. sapiens* caused the extinction of *H. neanderthalensis* – through competition for food, or shelter sites, or perhaps in outright warfare. But, equally, it has long been wondered whether modern humans interbred with Neanderthals and simply swamped them out genetically. DNA has been extracted from Neanderthal bones and, so far, analyses have turned up no evidence of interbreeding between the two. But many authorities believe that there must have been such interbreeding, even if there are no living descendants of such crosses.

Out of Africa

When the Roman historian Pliny the Elder (copying Aristotle) wrote '*ex Africa semper aliquid novi*' (there is always something new out of Africa), he cannot have dreamt how apposite this would be.

Because of the intense cultural overload, the study of human fossils has been beset by an even greater load of theoretical positions, usually held without or despite empirical data, than other areas of palaeontology. Human studies are rife with the search for ancestors, every new discovery tending to be assigned to an ancestral position in the main line to modern man instead of the more likely cousinly dead-end. It was long held, for example, that no more than one hominid species would (could) exist at any one time. Palaeo-anthropologists have also clung to various theories about the order of change among locomotor, dental, brain, and cultural adaptations, and about the patterns of hominid radiations in time and space. The latest theory to dominate the study of human origins proposes that Africa was at the centre of all human revolutionary radiations. And indeed, the one thing that all genera of human fossils have in common is Africa. No human fossils between the ages of 5 myr ago to 2 myr ago are known outside of that continent. Yet.

With human fossils, we have the opportunity to look at evolution on a very fine scale and to hypothesize about population migrations. The popular 'out of Africa' theory posits that Africa was the source of all the species that later colonized Eurasia, starting with *Homo erectus*, which had spread out of Africa into Asia by 2 myr ago and into Europe by 1.3 myr ago. Fossil evidence states that *Homo sapiens* arose some 250,000 yr ago, but its exodus from Africa occurred no earlier than about 100,000 years ago, and we had not colonized the entire Old World until 35,000 years ago. For comparison, the latest dates given by genetic comparisons of modern African and non-African populations give an age for the last common ancestor of all non-African *H. sapiens* of some 52,000 years, but with an experimental error of plus or minus 21,000 years (there is a mismatch here, as *H. sapiens* had crossed the oceanic barrier to Australia by some 60,000 years ago). There is also genetic evidence that initial population numbers in the first non-African populations were very small, going through a bottleneck on origin and again at about 35,000 years, which again neatly corresponds with the step in brain size increase and the advent of complex cultures. Reduction to very small population numbers will have increased the level of inbreeding, and this may possibly explain the high incidence of generic-based disease in living humans.

Homo heidelbergensis was in Europe and Asia by 500 kyr ago and *H. neanderthalensis* was in Europe at least 50,000 years before *H. sapiens*. In the 'out of Africa' theory, *Homo sapiens* caused the extinction of these other humans across the whole of its range, without any interbreeding. In this model, the modern races of humans have become differentiated only in the last 60,000 years or so. The alternative view is the 'multiregional continuity model', which posits that *H. erectus* was broadly transformed to *H. sapiens* across its entire range. *H. erectus* having diversified racially, the present races of *H. sapiens* developed in parallel from those older

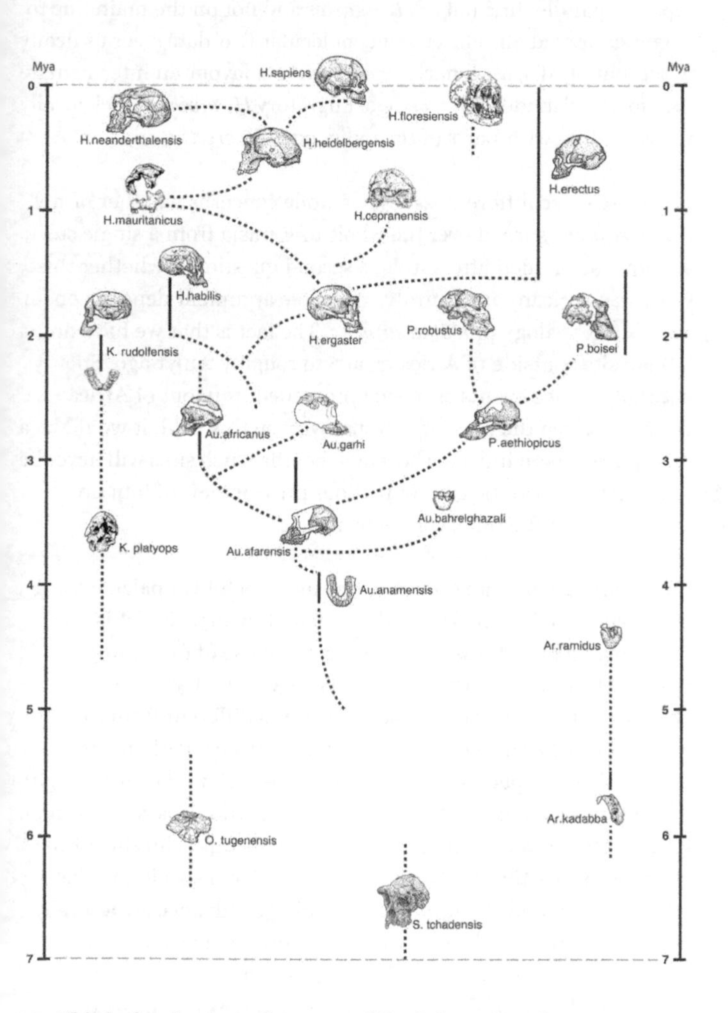

20. This is the simplest of many possible phylogenies of *Australopithecus*, *Paranthropus*, and *Homo*. We may confidently expect that this bare 'tree' will soon be expanded to a dense, leafy growth

diversifications. This is a debate that may not be soluble through the fossils alone unless *Homo erectus* can be shown to belong on a separate parallel line out of *H. ergaster* and not on the main line to *Homo sapiens* at all. The current 'molecular Eve' data steer us firmly to the 'out of Africa' scenario. Some workers favour an intermediate 'assimilation' model, with a single migratory *H. sapiens* stock locally interbreeding with parts of the indigenous *H. erectus*.

There are three different issues here: one concerns whether or not *Homo sapiens* spread over the whole of Eurasia from a single stock and only diversified afterwards; a second question is whether this stock was African; and, thirdly, the whole argument depends on an accurate genealogy of *Homo sapiens*. The fact is that we have no human sites outside of Africa from 5 to roughly 2 myr ago. So naturally we see events as having proceeded from out of Africa. However, given the pace of new discovery in this field, it would be a very brave person indeed who would bet that such sites will never be found. And we can be certain that our present view of human ancestry is not the final, complete one.

Fifty years ago, human evolution was the stepchild of palaeontology, bedevilled by a lack of data and a surfeit of theory. Now it is becoming an object lesson in the combined use of fossils and molecules in the reconstruction of phylogeny. Perhaps the evolutionary history of our own species was different from that of other species because of the factor of intelligence. Perhaps the origins of other species would look like this if viewed over the right timeframe – a complex story involving populations passing through severe bottlenecks, yet migrating over vast distances in short time spans; a species that perhaps assimilated related species, perhaps totally annihilated them, while continuing to advance structurally and behaviourally. All this is set in a context of environmental cycling between extremes, with advances and retreats of the ice and correlated closing and opening of lands and seas for subsistence and migration.

Chapter 9
Fakes and fortunes

As with postage stamps, coins, and Old Master paintings, the best fossils are rare and prohibitively costly. The important ones – the fossils that, like great works of art, break new ground – are few in any case, but perfectly preserved ones are even rarer in comparison with incomplete and broken ones. At the other end of the scale, the seemingly limitless supply of fishes and other fossils from the Eocene Green River Formation of Wyoming used to be available for the proverbial song. But no more: all fossils are now big business. Prices have risen dramatically in the last two decades due the immense popularity of palaeontology and a growing realization of the attractiveness of a well-prepared and displayed fossil. A poor ammonite from Morocco is today offered on a stall at the Oxford market for £75 and in New York City for $150. An 18-inch *Diplomystus* (a herring relative) from the Green River Eocene of Wyoming recently sold for $2,000 (and yet *Diplomystus* from Green River are common).

Mary Anning sold her first ichthyosaur from Lyme Regis for £23 (around £1,400 in today's money) in 1812; ten years later she sold a plesiosaur for £157 (equivalent today to some £9,000). The British Museum *Archaeopteryx* specimen of 1861 cost £700 (today, £35,000). O. C. Marsh paid US$1,000 in 1873 (today $15,000) for a pterodactyl from Eichstatt, a perfect *Rhamphorhynchus*, showing the membrane of the wing. All those numbers today would have to be multiplied another 10 to 20 times. A new *Archaeopteryx* specimen would certainly sell for at least half a million pounds. The record price for any fossil is the $8.4 million paid by a consortium including the Field Museum of Natural History in Chicago for the almost complete *Tyrannosaurus rex* skeleton 'Sue' in 1997. Fossils also have immense value in the academic sphere. A single really important fossil discovery could make the difference in a person's academic career and therefore be worth, over 20 years, many hundreds of thousands of pounds in salary and benefits.

Quite apart from the fiscal angle, there has always been an uneasy relationship between collectors and academics. Once again, it goes back at least to Mary Anning, when scholars competed with wealthy amateur collectors to buy her best specimens. Academics, who do not have the financial resources of some serious private individuals, always worry that the very best specimens found by commercial collectors will never see the scientific light of day, but will be hidden away in the private cabinets of eccentric millionaires. Commercial collectors (and no doubt some eccentric millionaires), however, see themselves as an important part of the palaeontological enterprise. Commercial collectors include some the most experienced field workers around. They are just as interested as scholars in making sure that important new finds are studied properly; but they also have to make a living.

Whether simply because they are a source of fame and fortune, or because fossils test our deep philosophical beliefs about the history of the universe and our place in it, it is no surprise that palaeontology has long been plagued by fakery. And some of these fakes have attracted an especially high degree of public interest. The lessons to be learned from these deceptions apply across all of science.

Piltdown

The best-known fraud in palaeontology concerns the most glamorous end of the business – human fossils. It is, of course, the

Piltdown forgery. The story is a masterpiece of deception. Charles Dawson was a country solicitor living in Lewes in Sussex, a keen amateur archaeologist and palaeontologist. In 1882, at the age of 18, he had presented an important collection of fossils to the British Museum (Natural History). Thirty years later, in February 1912, he wrote to Arthur Smith Woodward, the Keeper of the Department of Geology at the Museum, saying that he had found a site with Pleistocene fossils in Sussex, including 'part of a thick human skull', and a month later he sent Woodward a hippopotamus molar that he said was from the same site. Then finally, on 24 May, he took some fragments of a human skull to Woodward. (It later turned out that Dawson had shown some of the material to amateur archaeologist friends as early as 1908.)

This was a sensational discovery. No Pleistocene human fossil had ever been found in Britain, whereas in Germany the Neanderthals had been known since 1856 and there had just been a superb discovery in the form of the Heidelberg jaw – *Homo heidelbergensis*.

The site itself was most unpromising – essentially just a 'borrow pit' where gravel was taken for maintaining the farm roads of Barkham Manor but, over that summer, Dawson and Woodward collected more spectacular material, including a broken lower jaw. In all, they collected a quite complete assemblage of human and mammal fossils together with some putative stone implements. Woodward's technician made a reconstruction of the skull, showing an animal with a long forehead and jutting (prognathus) jaws but clearly human. In perfect accord with contemporary theory, Dawson's fossil was a combination of a relatively advanced cranium with a relatively primitive jaw.

So far, the only people who knew about the find were Dawson, Woodward, a few Museum staff, and a young French Jesuit seminary student. A keen palaeontologist, Pierre Teilhard de Chardin, later to become one of the discoverers of Peking Man and a major Catholic humanist philosopher, had been befriended by

21. Smith Woodward's reconstruction of the 'Piltdown skull' showed a relatively modern cranium with a primitive lower jaw and pronounced lower canine tooth

Dawson in 1909. He accompanied Dawson and Woodward for some of the digs in 1912 and again in 1913.

After the story was leaked to the *Manchester Guardian* newspaper, a proper announcement was made at a meeting of the Geological Society of London in December 1912. Cast copies of the skull were made available to suitable colleagues for study, and a fierce dispute immediately arose between Woodward and Arthur Keith of the Royal College of Surgeons. Ostensibly this was an academic disagreement over the reconstruction of a fossil but, in fact, it was also a dispute for 'ownership' of early man in Britain. Keith showed that cranial fragments could be fitted together as an essentially modern human skull. Much then would depend on the lower jaw, which was quite ape-like but lacked the essential condylar (joint) region and symphysis (chin) that would have been fully diagnostic. The lower canine teeth, if only they could be found, then became critical. If Woodward was right, then there should be a big projecting canine as in an ape; if Keith's reconstruction was right, it would be smaller, as in modern humans.

On 30 August 1913, Teilhard de Chardin picked up a canine at the Piltdown site. It exactly fitted the Woodward reconstruction. Keith admitted defeat. But the debate over Piltdown would not go away, even when Dawson announced a new find from nearby Barkham Mills. Nor when he announced material from a third site, at Sheffield Park. Finally, one of the most bizarre discoveries from Piltdown came in 1914 when Woodward picked up a large, blade-shaped implement made from the scapula of an elephant. The 'first Englishman' as Woodward had started calling the find, now had a unique bone tool like nothing seen anywhere in the world.

As early as 1913, William King Gregory of the American Museum of Natural History had warned that the whole thing might be a 'deliberate hoax . . . a Negro or Australian skull and a broken ape jaw'. But even he was later converted to the Woodward view by the weight of evidence. In 1916, Dawson died and the discoveries stopped.

After his retirement, Woodward went to live near Piltdown and continued to search, but not the slightest trace of bone or of a stone implement ever appeared again. By the 1940s, as theories changed, Piltdown man had become a monstrous anomaly: it should have been the other way around – a more modern jaw, an archaic cranium. In 1953, chemical analyses revealed that it was a fake. The culprit was apparently Dawson; he had set up Woodward brilliantly, teasing him along with small nuggets of information and then allowing him to be part of the major discoveries. The young priest Teilhard had been sucked in for an extra element of authenticity. Every time doubts were expressed, a bit of new evidence was conveniently produced.

Piltdown is useful because it shows why people commit such frauds and why they succeed. Dawson gained acceptance as a serious scientist and would certainly have been elected Fellow of the Royal Society had he not died. Woodward did become a famous scientist and the status of the British Museum (Natural History) was enhanced. Moreover, politically, England had needed an early human fossil to counter the discovery from Heidelberg. Dawson provided it; he created a composite of a partial human skull with an ape's jaw, and doctored it all to look old.

The story has a final twist. For thirty years, the whole scientific establishment had rallied behind the so-called *Eohomo dawsoni*. That had made it very difficult to challenge Woodward or Dawson. Yet it seems possible that someone did. Several old Museum staff members even said in their later years that someone in the Museum had been involved. Joseph Weiner at Oxford, who exposed the fraud in 1953, noted that the famous canine tooth had not been elaborately treated chemically to age it in the same way as the other material: it was merely painted with artist's oil paint. And the bone tool was absurd. The possibility exists that one or more of the younger Museum staff realized that Piltdown was a fake and decided to expose it to the principals by planting their own fakes. The first of these was the canine. But the scientific establishment could not see beyond using it to distinguish between two interpretations of what was a fake anyway, and their theories said that *it should have been true*. Then the new hoaxers tried again. The shape of the bone implement gives it all away: they gave the 'first Englishman' his own first cricket bat! However, when Dawson died, the hoaxers were left in the awkward position as the only living people who had contributed to the fraud, so now they had to lie low. One of them has now almost certainly been identified as Martin Hinton, then a junior volunteer at the Museum hoping to land a permanent job. He may even have been aided by Teilhard de Chardin, for it was he who 'found' the canine, and he would have been furious at having been duped. Some people think that Hinton was responsible for the whole fraud, but Dawson turns out to have had a ‘record’. He had created previous fakes in archaeology, with the identical *modus operandi* of producing new evidence just when things were going badly for him. I still see Dawson as the instigator of the fraud.

All in all, Piltdown is a sorry tale – not just for Dawson’s wickedness, but for the gullibility and arrogance of the scientific establishment and their ruthlessness in putting down dissenters.

Lugensteine – ‘lying stones’

Perhaps the first scholar to fall foul of the problem of faked fossils was someone whose imaginative theories about the causes of fossils might otherwise have made him an early patron saint of palaeontology. Johan Bartholomew Adam Beringer (1667–1740) was Dean of the Medical Faculty and Court Physician at Würzburg, Germany, and a keen geologist. At nearby quarries on Mount Eivelstadt in the 1720s, he had found a range of familiar fossil shells and ammonites and had begun to question their causes, tending first to support the theory that they were the result of the Deluge. A busy man, he entrusted further collecting to three assistants.

In 1725 some unusual fossils were brought to him – they appeared to be of crayfish, worms, frogs, and plants. It seemed to Beringer that: ‘In this one place, as in a full horn of plenty, all those things are gathered that Nature has divided up among the pits, caverns and hiding places of other provinces.’ As more and more treasures emerged, he began to write a great book about these fossils and even conducted expeditions to the quarry for colleagues to find their own specimens. However, two colleagues in the University soon spread the word that Beringer’s fossils were fakes. Beringer dismissed these claims and, in the meantime, the finds became even more bizarre and thrilling: ‘clear images of the sun and the moon, of stars, and of comets radiant with flaming tail’. Finally, the assistants brought him ‘splendid tablets . . . marked with the ineffable name of the Divine Jehovah in characters of Latin, Greek and Hebrew’.

Instead of coming to his senses, Beringer was totally taken in. He began to theorize about new kinds of causal factors in the heavens. He decided that his fossils were not the same as other 'formed stones' stemming from the Great Flood or 'from the seminal and formative power of a generative breath or subterranean Archaeus or Panspermia'. Nor were they 'chance products of our wonderful mountain'. Instead, he developed an elaborate theory based on the properties of light: 'a flow of minute solar particles which being of fiery essence pass through the atmosphere . . . (with) the truly marvellous faculty of depicting, portraying and forming the images of the bodies that it falls on in its flow'. Therefore, he asked, could it not be supposed that it has a 'certain active and creative power of imprinting on suitable matter the same forms of which it has already taken the impression'? In other words, without being the slightest bit flippant, it seems that he was hypothesizing a sort of early version of a xerography process. The light would form images of living creatures or of the words on tombstones, and transfer them to 'mud, clay, sand and soft stones'.

Then the bubble burst. An inquiry was called, revealing that Beringer had been the innocent (if all too gullible) victim of a most wicked fraud. His first fossils had been real, all the rest were fakes. Even worse, they were not put there by students as a practical joke, but by his own professional colleagues. They had decided to discredit him. One of his 'helpers', a youth named Christian Zanger, had carved and planted the forgeries for the others to take to Beringer. But Zanger had been paid by none other than Ignatz Roderich, the Professor of Geography, Algebra and Analysis at Würzburg University. Georg von Ekhart, Privy Councillor and Librarian to the Court and to the University, had helped polish them. A truly vicious little story of academic rivalries.

22. Among the fanciful 'fossil' objects carved by rivals to deceive Beringer were a sort of crayfish (or perhaps it is a slug) and hieroglyphs

Archaeoraptor was a fake: what about *Archaeopteryx*?

Nothing delights us more than conspiracy theories and *schadenfreude* – seeing the great brought low. In 1985, the astronomer Fred Hoyle and his associate N. Wickramasinghe challenged the integrity of no less a fossil than that totem of palaeontological totems: *Archaeopteryx* itself.

Eight specimens of *Archaeopteryx* are known, the first being a single feather discovered in 1860. In 1861, a superb specimen came to light at the Solnhofen lithographic limestone quarries in Bavaria and was sold to the British Museum in London. In 1877, an equally spectacular specimen was found at Eichstatt and sold to the Humboldt Museum in Berlin. A specimen that had been collected even earlier was eventually found languishing in a museum drawer in Munich having been misclassified as a pterodactyl. The most recent new discovery was in 1960. *Archaeopteryx* combines a number of reptilian features (teeth rather than a horny bill; a long bony tail; the trunk vertebrae are not fused, nor are the bones of the hand) and bird features (feathers and a furcula, or wishbone). *Archaeopteryx* is a 'missing link' from Late Jurassic times, closer to being a small coelosaurian dinosaur than a bird, but with true feathers.

On the basis of some rather crude photography, Hoyle and Wickramasinghe claimed that the impressions of feathers on the British Museum specimen were forged by taking a slurry of the original limestone and adhesive and pressing modern feathers into it. The Natural History Museum (as it is now called) had to expend a vast amount of staff time and energy on demonstrating what was obvious to scholars all along – that *Archaeopteryx* is genuine. Hoyle was an extremely distinguished scientist and it is not clear why he started this particular investigation, but his charge of forgery gathered huge momentum. It is consistent, however, with his pugnacious personality. In astronomy, he is well known for challenging the notion of the Universe originating in a Big Bang. Because *Archaeopteryx* is the poster-boy for fossils and evolution, anti-evolutionists seized with relish upon the charge that it was a fake, and they have proved happy to ignore the evidence that it is real.

Unfortunately, an even better example for their cause was provided by the supposed half-bird/half-dinosaur *Archaeoraptor*, from the Cretaceous of China. A particularly 'hot' topic in palaeontology in recent years has concerned the origins of birds: from which branch of reptiles are they descended? Many believe that the closest relatives of birds are the dinosaurs and that, in fact, dinosaurs did not become extinct at the end of the Cretaceous but persist today in the form of birds. In that case, there might exist cases of true dinosaurs with feathers, but not yet the other characteristics of birds. Almost inevitably, in the year 2000, such an eagerly sought beast appeared: a dromeosaurid dinosaur but with real feathers, to which the name *Archaeoraptor* was given ('raptor' – in the sense of a bird of prey – being a particularly charged word in this field). This was a slab of fossil from a locality that has produced many interesting examples of early true birds. It was sold for some $80,000 to an amateur collector in the United States and seemed to be an even better 'missing link' than *Archaeopteryx*.

Most serious scientists, however, were wary of the 'find', and it was never subject to scientific peer review for a mainstream publication. Instead, it was announced in a long article in *National Geographic Magazine*. Within a year, it was revealed to be a composite of several different slabs from the same site, skilfully put together. It does have feathers but only because that part came from a known fossil bird – *Yanornis*. It was a dinosaur because the other parts came from several different slabs, each with bits of dinosaur on them.

The motive for this forgery on the part of its fabricators in China was simple profit. But once again its acceptance, limited as it was,

23. Literally half a bird fossil and half a dinosaur, *Archaeoraptor* did not fool everyone

depended on there being people who really wished it to be 'true'. They were trapped by the conviction that such a creature should have existed and, no doubt, by the rush to be first and most famous in the discovery of bird ancestors. Much credit goes to the Chinese scientists who faithfully revealed the truth about the forgery.

Archaeoraptor was not the first fossil to be 'embellished' to improve its sales value. Those who bought and studied it will not be the last to be fooled, or to fool themselves. More convincing cases of dinosaurs with at least a sort of fuzzy proto-feather have been found since the *Archaeoraptor* debacle, so the scientific impact of the fraud has been negligible. However, the damage done to the reputation of evolutionary palaeontology by this fraud and the libel of *Archaeopteryx* has been immense. As even a cursory search on the Internet reveals, they have given the less scrupulous elements of the creation–intelligent-design movement much ammunition for criticizing and dismissing other areas of evolution and palaeontology.

Chapter 10
Back to the future

If we look back at how our understanding of fossils and earth history has changed over time, we can try to guess how it might change again in the future. It has famously been said that there are more scientists alive today than all those who have lived previously combined. The same is true for fossils: more have been collected in the past 50 years than in all of previous history. The rate of discovery will level off eventually. Meanwhile, judging by the pattern of discovery in the past 25 years, one area of study in which new discoveries will transform our understanding is surely the first 2.5 byr of earth history and the crucial episodes in late Precambrian time when modern kinds of animals and plants first arose. We can expect to improve estimates of the course of diversity change through the Phanerozoic and to refine our knowledge of the causes of both mass extinctions and mass diversifications of organisms. But a palaeontological version of Murphy's Law will surely still plague us. In all probability, in the future as now, new fossils will be more likely to extend the range of particular groups deeper into time than to bring them closer to other groups. And, for every gap in the history of life that is filled, a new, different gap will be defined. Even as we find more human fossils, e. e. cummings's observation will apply: 'always the more beautiful answer who asks the more beautiful question'.

So far, our more general theories about fossils have turned out to be quite robust – while the details have changed. Even so, it is hard to be sure to what our extent our interpretations of the fossil record are vulnerable to the charge of being overly theory-laden. Can we imagine a different theoretical framework that would cause us to revise our ideas? For example, if history had been upside down, and all the early intellectuals and their discoveries had been in China and Australia, or all made by Buddhists, how would our view of fossils and history be different from the present Eurocentric version?

Questions for the future might include: will the present high rates of extinction of megafauna stand out in the future record, or merely register as noise within some larger signal – an interglacial blip in the course of fossil history? Ten myr from now, perhaps sooner, there will be an answer to this, one of the more interesting aspects of the Pleistocene and Holocene record. So far, it seems that the last 2-myr period – a time of great environmental stress and huge shifts in the home ranges of species (including humans) – has been a time of extinctions rather than the focus for formation of new species. (It would be perfectly ironic if the age that discovered a viable theory of evolution was also an age of extremely slow rates of origins of new species, making empirical confirmation of the theory even more difficult.) But is all this a correct reading of the record, and, if so, how long will the pattern continue before originations begin to surge again? What will be the trigger? And what will turn out to have been the 'next big thing' on the evolutionary stage – another species of human, a better cockroach, or yet another virus?

The question 'what comes next?' is closely bound up with a much deeper question about the course of past evolutionary history: was there something inherent in the pattern of evolutionary change that produced modern floras and faunas, including, of course, ourselves? If so, then if we were somehow to wind the clock back to, say, Permian times and then let it run forwards again, we would finish up where we are now. The view that the result would be the same is favoured by those who see history as end-directed and the purpose of evolutionary change all along to have been to produce humans (and presumably rattlesnakes, snails, and the HIV virus). The contrary view is that the chances of evolution running exactly the same course would be infinitesimally small. Evolution would, by chance, produce an entirely different set of organisms; birds and mammals as we know them, let alone primates and humans, might never have appeared.

Setting aside the religious consideration of whether there is some non-material cause of the universe, the scientific answer to this conundrum is that both views of history are partially correct. Evolution is driven by chance. But what happens by chance at each instant of time is predicated in large part by the concrete fact of what then exists and what has been before, especially in the evolving cascades of genetic and developmental processes that produce an adult out of a single egg. Thus humans did not evolve from spiders, and the next million years of evolution will not produce humans with butterfly wings, or insects with mammary glands.

If there were no chance in the equation, we could predict what the inhabitants of the world ten million years from now would look like. And, of course, if the whole purpose of earth history and evolution has been to produce modern humans, then in ten million years the new fossil record would necessarily show no change at all: evolution would have stopped because its goals had been achieved. As we can already see that humans have changed quite a lot in the last million years, however, the likelihood of further change seems to be 100 per cent.

The formation of fossils and creation of layers of rock is not something that happened only in the past. Today, all around us, new potential fossils are being laid down in sediment; just under the surface are the evidences of similar processes that started a thousand, a hundred thousand, and a million years ago. And the process continues. Indeed, will current climate trends and environmental uses produce an unusually high level of erosion and sedimentation to contain those fossils?

Which of today's flora and fauna will likely be preserved as fossils? As an old joke has it, the strata being laid down today will eventually be named the Dustbinian Formation, with its accumulations of the flotsam and jetsam of our throwaway society. Its signature fossils will be a concentration of macroscopic and microscopic pellets of styrofoam and other plastics – they are already showing up in marine sediments all over the globe. If we wonder where all the iridium came from that marks sediments laid down at the time of the Cretaceous–Tertiary boundary (and extinctions), our successors will be equally puzzled by the anomalous levels of Holocene iron, aluminium, and glass, and unusually high levels of radioactive materials. Future archaeologists, and whatever creatures inherit the earth after we are all extinct, will find that the commonest Dustbinian fossils are domestic animals like chickens, sheep, and cows, with horses, cats, and dogs following close behind. In a dramatic reversal from pre-Holocene deposits, remains of humans will be abundant – all evidence of a massive post-glacial population explosion. As for extinctions, sophisticated explanations will be sought for the fact that many of the most prominent extinctions were of the larger animals, but the correlation between the rise of human population numbers and the other major shifts in diversity will be overwhelming.

Given our penchant for burying things in holes in the ground (not just each other but, for example, the domestic waste rubbish we bury in Pleistocene sand and gravel pits, Carboniferous limestone quarries, and mines of all ages), the remains of our culture will turn up in some odd geological formations. Future stratigraphers will also discover that we have moved vast amounts of rock from one place to another – and fossils with them.

But some of the most difficult puzzles for our successors to work out will concern the distribution of animals and plants. From deposits dated from about 200 years ago, they will find that large numbers of organisms – from grasses to horses – have suddenly extended their ranges. How will they account for the seemingly miraculous appearance of European rabbits and the house sparrow in Australia, or of maize in Europe? Occasionally, future palaeontologists will turn up something truly baffling, like the skeleton of a zoo animal – an Indian elephant in Mexico or a tapir in France, tigers in Australia, or koalas in Russia. How could it be, they might wonder, that remains of a supremely rare animal like the giant panda could be all over the world, just before the species became extinct?

Horses will be a particularly interesting puzzle to work out in the future because the fossil record of the modern horse will be extensive. The present record shows that the horse became extinct in North America about 10,000 years ago, and was never in South America. It was re-introduced to North America by the Spanish, and by 200 years ago was flourishing wild on both continents. Will this introduction be discernible in the fossil record, or will the 10,000-year gap between the last native horses in North America and the first introductions simply be dismissed as a technical artefact in an otherwise continuous record?

On a positive note, 50 myr from now, the huge deposits of organic matter currently being deposited (rather problematically) as garbage and sewage sludge in our river valleys and around our coasts will have created a new set of coal, gas, and oil reserves. Paper alone will account for large hydrocarbon deposits. What is unknown is how thick all the deposits will be: how long the Dustbinian will persist. A great deal may depend on the glacial history of the earth. A lot of the surface Tertiary record is missing from Europe and North America, having been swept away by the Pleistocene glaciations. If the glaciers return in force, 100,000 years from now the buildings and other evidences of civilization in

Europe and North America will have been ground back to a mud rich in carbon, plastic, and metal and redeposited elsewhere, mostly in the seas. Whoever then rediscovers Palaeolithic cave paintings might ask: whatever happened to those people?

Further reading

Anna K. Behrensmeyer and Andrew P. Hill, *Fossils in the Making* (University of Chicago Press, 1980)
David J. Bottjer, Walter Etter, James W. Haagdorn, and Carol M. Tang, *Exceptional Fossil Preservation* (Columbia University Press, 2001)
Peter J. Bowler, *Evolution: The History of an Idea*, 3rd edn. (University of California Press, 2003)
D. E. G. Briggs and P. R. Crowther, *Palaeobiology II* (Blackwell Publishing, 2003)
Eric Buffeteau, *A Short History of Vertebrate Palaeontology* (Croom Helm, 1987)
Deborah Cadbury, *The Dinosaur Hunters* (Fourth Estate, 2000)
Edwin H. Colbert, *Men and Dinosaurs* (Dutton, 1966)
A. Cutler, *The Seashell on the Mountain Top* (Dutton, 2003)
Richard Fortey, *Life: An Unauthorised Biography* (HarperCollins, 1997)
— *Trilobite* (HarperCollins, 2000)
John C. Greene, *The Death of Adam* (Iowa State University Press, 1959)
Mark Jaffe, *The Gilded Dinosaur* (Crown, 2000)
Melvin E. Jahn and Daniel J. Woolf, *The Lying Stones of Dr Beringer* (University of California Press, 1963)
T. S. Kemp, *Fossils and Evolution* (Oxford University Press, 1999)
Arthur O. Lovejoy, *The Great Chain of Being* (Harvard University Press, 1939)
W. J. T. Mitchell, *The Last Dinosaur Book* (University of Chicago Press, 1998)

Donald Prothero, *Bringing Fossils to Life* (WCB McGraw-Hill, 1998)
David M. Raup and Steven M. Stanley, *Principles of Paleontology* (Freeman, 1971)
Martin Redfern, *The Earth* (Oxford University Press, 2003)
Martin J. S. Rudwick, *The Meaning of Fossils*, 2nd edn. (Science History Publications, 1976)
— *Scenes from Deep Time* (University of Chicago Press, 1992)
George Gaylord Simpson, *Tempo and Mode in Evolution* (Columbia University Press, 1944)
Ian Tattersall, *The Fossil Trail* (Oxford University Press, 1995)
Keith S. Thomson, *Living Fossil* (W. W. Norton, 1991)

Index

D

E

F

G

H

I

J

K

L

M

N